Nature and Experience

Nature and Experience

Phenomenology and the Environment

Edited by Bryan E. Bannon

ROWMAN &
LITTLEFIELD
INTERNATIONAL
London • New York

Published by Rowman & Littlefield International, Ltd.
Unit A, Whitacre Mews, 26-34 Stannary Street, London SE11 4AB
www.rowmaninternational.com

Rowman & Littlefield International, Ltd. is an affiliate of Rowman & Littlefield
4501 Forbes Boulevard, Suite 200, Lanham, Maryland 20706, USA
With additional offices in Boulder, New York, Toronto (Canada), and London (UK)
www.rowman.com

British Library Cataloguing in Publication Information Available
A catalogue record for this book is available from the British Library

ISBN: HB 978-1-7834-8520-8
ISBN: PB 978-1-7834-8521-5

Library of Congress Cataloging-in-Publication Data

Name: Bannon, Bryan E., 1979-editor.
Title: Nature and experience : phenomenology and the environment : edited by Bryan E. Bannon.
Description: Lanham : Rowman & Littlefield International, 2016. | Includes bibliographical references and index.
Identifiers: LCCN 2016012537 (print) | LCCN 2016013196 (ebook) | ISBN 9781783485208 (cloth : alkaline paper) | ISBN 9781783485215 (pbk. : alk. paper) | ISBN 9781783485222 (electronic)
Subjects: LCSH: Philosophy of nature. | Ecology. | Phenomenology.
Classification: LCC BD581 .N36978 2016 (print) | LCC BD581 (ebook) | DDC 113--dc23 | LC record available at http://lccn.loc.gov/2016012537

∞™ The paper used in this publication meets the minimum requirements of American National Standard for Information Sciences Permanence of Paper for Printed Library Materials, ANSI/NISO Z39.48-1992.

Printed in the United States of America

Contents

Introduction

Bryan E. Bannon

Speaking of 'nature' conjures up images of the world 'out there', water and sky, tree and stone. Conversely, speaking of 'experience' recalls the world 'in here', thought and feeling, values and memory. Hence, encountering a book called *Nature and Experience*, one might expect to read of how the world out there relates to the world in here, how, for example, the objective properties of the world are qualitatively experienced by a conscious perceiver or how an intelligent subject constructs around itself a world of meaning. While these are fair expectations, for those of us who have adopted phenomenology as a philosophical method the terms 'nature' and 'experience' do not refer to two different kinds of things, objects and consciousness, but to a specific set of questions and problems that emerge when one no longer takes for granted that the world can be divided up into such categories. For this reason, the chapters contained herein can be thought of as an attempt to grapple with one set of these questions, those that pertain to the practice of environmental philosophy.

Typically, environmental philosophy is thought to be synonymous with environmental ethics, but, as the reader will see, the term also possesses a much broader meaning for the phenomenologist. Environmental philosophy is an attempt both to return to an older, more speculative form of philosophy wherein metaphysical questions concerning nature can be asked without embarrassment and to analyse the manner in which our contemporary relationship to the environing world alters the traditional understanding of philosophical problems and concepts. In this way, environmental philosophy seeks to project a different possible future for both philosophy and the world around us. The environmental philosopher, then, asks the traditional philosophical questions while remaining anchored within a concern for and attentiveness to the natural environment.

While phenomenologists by no means possess a monopoly upon the practice of environmental philosophy, many of us believe that the method holds special promise with respect to its concerns. Because the nature of nature has from its inception been a central question of phenomenological inquiry, phenomenologists have discussed and refined responses to many of the questions just arising in English-language philosophy for decades. Saying so, of course, is not to say that the phenomenological tradition possesses fixed *answers* to these questions, especially when one considers that two of the hallmarks of phenomenological philosophy are the emphasis placed upon interpretation and the open-endedness of its results. Rather, the thought is that phenomenology can contribute the hard-won insights into what may or may not be valuable lines of inquiry, styles of asking a question or ways of responding to a question in order to avoid past mistakes. One of the purposes of this volume is to make such contributions more apparent to those outside the phenomenological community. To this end, the chapters collected here address topics of interest within the environmental humanities more generally (e.g., how best to approach the concept of nature given contemporary scepticism toward the concept, whether and in what way we can discuss the agency of nature, the role metaphor and narratives play in shaping the human relationship with nature, the human aesthetic connection to the environment, etc.). Moreover, they do so, generally, without the detailed examination and analysis of the texts of specific historical figures that frequently make phenomenological philosophy so inaccessible to those who have not studied those texts. Hence, readers are not expected to have any previous experience with phenomenology and its leading lights in order to engage with the chapters here.

To make this engagement between practitioner and non-practitioner possible, it may prove helpful to say a few words regarding what phenomenology is and what its basic concepts are prior to proceeding. At its most fundamental, phenomenology is a philosophical method that begins with an examination of experience. Rather than an empirical conception of experience, however, that attempts to reduce experience to its simplest terms in order to understand it, phenomenology seeks to understand experience in its complexity. In other words, phenomenology eschews idealized and abstract models of experience in favour of engaging the world as we actually experience it. Even though individual phenomenologists will offer distinct accounts of what experience is and its meaning, one can safely assert that phenomenology begins within the everyday experience of the world and seeks to understand the structure of that experience.

Because phenomenology begins in and reckons with experience, the concept itself is one of the most debated within the field. Most phenomenologists take experience to be characterized by what are termed 'intentional relationships' or 'intentionality'. Intentionality, conceived as a relationship, has little to do with the dominant sense of the term within the philosophy of mind, roughly, the mind's power to represent the world to itself. Rather, the phenomenological conception of intentionality is traditionally expressed using Edmund Husserl's phrase 'consciousness is consciousness of', which, admittedly, can readily be misunderstood to be referring to the mind's power to represent. Another way to understand the formulation, however, would be to understand 'the mind' not as something existing independently of the objective world whose relationship to it needs to be explained, but as a particular way of engaging with the surrounding world. More simply, consciousness in not a *thing* or a *substance*; it is a myriad of relationships between a body and its environs, the varieties of which call for description and analysis.

Because consciousness is a relationship between a perceived 'I' and its surroundings, the phenomenologist will frequently speak of the 'constitution' of an object or of consciousness. This term refers to the manner in which the intentional object, that is, that of which one is conscious, and one's conscious awareness mutually influence one another. For example, consider one's relationship to food. It seems to be a universal feature of the human experience to differentiate between what can be eaten and what can't. We typically don't need to think about this distinction as it happens immediately within perception: some things just *are* food and some things *aren't*. Seeing something *as* something due to the kind of relationship one has to the thing is half of what it means to be constituted. The other half pertains to the person doing the perceiving: To the extent that one views something as food, one becomes a certain kind of person (even if the kind of person one becomes is something as simple as 'the person who eats X'). Constitution, then, is a two-sided phenomenon; relations are constitutive of what each party of the relationship is.

If the story concluded with these constitutive relations, it would be fair to criticize phenomenology for being a subjectivist philosophy. However, when one begins to question *why* things are constituted the way they are, we move beyond the individual subject, revealing what phenomenologists tend to identify with the intersubjective dimension of experience, the 'world' or 'lifeworld'. The concept of world refers to the manner in which our surroundings make sense to us, appearing as a system of related things. These apparent relations are not arbitrary or subjective, however; they are a product of both the historical legacy of human interpretation and the manner in which the world shapes and moulds our interpretations through its recalcitrance and compliance. Especially when one takes into account how individuals emerge from a process of physical and cognitive development, it should be clear that this broader domain is primary and is what generates the initial shape of experience that we later identify as our own and with which we engage critically. In this way, the intersubjective lifeworld is given in advance of our existence as an individual, and it is through the lifeworld that things take on the particular meanings that they do, for example, that kale is for eating and wood chips are not. Obviously, the lifeworld will be specific to geographical and historical variations in human existence.

Normally, we go about our day-to-day business uncritically relying upon the relations we perceive within the lifeworld. Phenomenologists call this manner of living the 'natural attitude'. Within the natural attitude, we take for granted the obviousness of these relations, the existence of others and the world, the validity of our projects, etc. In other words, the natural attitude, for better or worse, reflects the meanings, values and prejudices of the received lifeworld. The natural attitude is not a bad thing per se, but phenomenological reflection requires that we step outside of it in order to reflect upon the very theoretical commitments in which it consists, for example, that one's mind is distinct from one's body, that consciousness is free and the material world is causally determined or that there is nothing odd about eating other animals. This process of setting aside the philosophical prejudices of the natural attitude is called 'bracketing' or the phenomenological epoché. The thought is that by stepping outside the natural attitude, by suspending one's commitments to the obviousness of those relations, one can place oneself in a position to examine how those commitments are relationally constituted, evaluate whether there might be other, more productive ways of understanding those relations, and, if so, begin reconfiguring those relations.

The account offered here is an extremely general one and is hardly exhaustive, nor, I imagine, would it be universally endorsed. I have offered it in the spirit of describing the project of phenomenology in as ecumenical a manner as possible. After all, as the chapters in this volume will bear out, there are myriad conceptions of what phenomenology is, what its appropriate philosophical goals are, and how its main concepts and methods function. Since the chapters here represent several different schools of phenomenology, rather than viewing them as a presentation of a definitive view regarding what all phenomenologists believe on a given topic, it will be more productive to view the chapters as a window into the kinds of concerns that occupy us.

The volume is divided into three distinct sections, the first of which pertains to the question of how phenomenologists think of nature itself. While each chapter will recommend a particular way of thinking about what nature is and the human relationship to it, there are nevertheless several themes that unite them all. Bryan Smyth's chapter takes the most general approach, arguing that nature is best understood in terms of a mythic structure: Our account of what nature is conditions all of our other understandings of what the world is like, and yet we can never offer an account of what nature is beyond question since every means we have for doing so takes for granted some conception of nature. Because nature is what Smyth calls the 'dynamic horizon of significance', that is, the concept that delimits the totality of reality, it is unavoidable that we provide some narrative regarding what nature is. Properly speaking, then, we are faced with the challenge of determining what the most appropriate mythic narrative concerning nature could be. The content of this narrative, moreover, is subject to normative evaluation since the concept can license social injustices. We must, in other words, assume responsibility for the political nature of nature.

The next two chapters, by Janet Donohoe and Tom Greaves, respectively, both offer an account of what that narrative should be like. Donohoe takes up the perspective of genetic phenomenology, that is, an inquiry into the layers of cultural and personal experience that condition our current understanding and experience of the world. In her view, like Smyth, nature serves as a horizon of experience, but she views nature in a more foundational sense: Nature is that which is pregiven, the framework for all experience. But what does it mean to be the framework for all experience? For Donohoe, this refers back to the structure of constitution in which meaningful world and perceiving person cannot exist as they are without one another. In this interconnection between world and its inhabitant, there is an interplay of relations, ranging from the physical feel of a particular place, to cultural interpretations of the environment and the manner in which one relates to that place, that we take to be normal. This experience of being at home in the world is the phenomenon of the homeworld, which comes to the fore of our attention only when we encounter the alienworld, that is, a world with divergent constituent relations. In confronting the alienworld, we come to understand that the homeworld can be experienced differently, which allows us to enter into a critical relationship with it.

Greaves takes a different approach to the phenomenology of nature, identifying nature not with a particular horizon, but as a qualitative dimension of experience. In order to understand nature, claims Greaves, phenomenologists must pay greater attention to the *intensity* of experience and not only its intentionality. Intensification occurs along two axes, the first being the manner in which phenomena overflow and exceed the bounds of ordinary experience and the second being the manner in which the things themselves carve out the space within our experience of them for their appearance. In this way, intensity is intrinsically linked to the concepts of birth and growth within a given experience. The similarity of Greaves's view with Smyth's and Donohoe's lies in their agreement that nature should not be considered to be a thing, which, as Greaves points out, is a longstanding philosophical prejudice. Conceiving of nature not as some thing or type of things, but as a quality of experience both enables us to avoid thinking of humanity and human artefacts as unnatural and helps to undermine the distinction between facts and values.

Simon P. James's chapter addresses the persistent criticism of phenomenology as being inherently anthropocentric. Within environmental ethics, this criticism tends to be a rather severe one, with the dominant view being that anthropocentric philosophies are to be avoided insofar as they have contributed greatly to our current environmental situation. While it is clear that most phenomenologists do not endorse anthropocentric values, that is, believe that the appropriate locus of value in the world is human interests, it nevertheless remains the case that phenomenology could fall prey to an *ontological* anthropocentrism. Briefly, ontological anthropocentrism is the view that everything that exists is what it is in virtue of its being present for a human being. The thought is that if a philosophy is ontologically anthropocentric, then it both enables the anthropocentrism of values and forecloses more effective and just means of addressing environmental problems. James argues against such a characterization of phenomenology by referring to the work of Maurice Merleau-Ponty and the manner in which the phenomenological approach reveals the two-sidedness of constitutive acts: When we engage with the world, we are directed toward the world in particular ways that shape our relations to the world precisely by the things with which we are engaged. In this way, it is not only that things bear the mark of the human, but also that humans carry within us the marks of the nonhuman. Hence, phenomenology is not ontologically anthropocentric because, we might say, *everything* is shaped in its relations to otherness.

The final chapter in this section, my own, takes up this relational picture of the world with respect to the question of the value of nature. Environmentalists have been arguing for decades that nature possesses some value independent of the value imparted to it by human beings. In my view, this way of asking the question occludes the fact that value is not a thing possessed but, as Greaves argues for nature in general, is a kind of qualitative dimension of experience itself. Thus, nature does not have inherent or intrinsic value, but the kinds of relations engendered within nature are themselves valuable. To see why this might be the case and its implications for environmental philosophy in general, I develop a more general conception of experience than the one described above, one that applies not just to human beings, but to embodied beings in general. In this way, we are called to recognize nature not as something possessing value or on which value is imparted, but as a domain that is constituted through the valuing relations of which it consists.

The second section takes up some of the perennial concerns of environmental humanities scholars: the ecological imagination, the agency of nature, the role metaphors play in shaping our relationship to the natural world, and why we are subject to moral failure. In many ways, the chapters in this section are simply raising questions, posing familiar problems in a different fashion, and proposing an approach to dealing with them.

Mark Thorsby, for example, takes up the question of moral responsibility with respect to the environment by discussing how such a pervasive moral failure is possible. Indeed, the question is even more properly understood in terms of why most people don't regard their individual actions as pertaining to the decline of the environment at all. The problem lies in the intersubjective lifeworld: The lifeworld is constituted through *practices*, which are taken for granted as appropriate. Because the lifeworld is constituted through practices, and practices always require and take place in a material world, the lifeworld is not *subjective*, nor is it an aggregate of subjective worlds. As Thorsby puts the point, intersubjectivity is a real feature of the world. Keep in mind as well Donohoe's and Smyth's discussions earlier, namely that practices are only ever intelligible upon the background of an environing world. In a sense, then, practices that destroy the environing world, quite literally in the case of nature, present us with a contradiction insofar as the practices that give meaning to life within the world are also the practices that destroy possibility of life in that world. Again like Donohoe and Smyth, Thorsby recommends a genetic phenomenological analysis to uncover the historical structures of the lifeworld that place us upon this self-destructive path. What we find is that the moral failure lies in our inability to imagine a different future for ourselves outside of our current lifeworld, constituted as it is by environmentally devastating practices.

But what is to be done? The next three chapters each present a distinct perspective on this point, not as a direct response to Thorsby, but as a way of approaching environmental issues more generally. The first is Elise Springer's chapter, 'Metaphor and Weather', in which she calls upon us to think more critically about metaphor and how metaphor relates to our ability to voice our concerns about the environment. We misunderstand metaphors when we view them as primarily cognitive structures because (1) they are not necessarily modelled on logical statements, (2) fail to recognize the reciprocal relation between the terms put in relation, (3) reduce them to internal mental processes (e.g., brain states or subjective association), or (4) understand them as tools rather than a means by which human expressive life occurs. A metaphor shifts our understanding, moves us toward perceiving differently and elicits responses. We do not merely *use* metaphors, but metaphors shape our practices and who we are because they are things we *undergo*. This Nietzschean language is no mistake: As for Nietzsche, Springer describes the experience of metaphor as bidirectional, drawing upon some source of meaning (the understood term) to cast the described term in a new light, but in the process of doing so altering our relation to the first. Since the utilization of metaphor is inevitable while trying to make sense of our environmental situation, we must exercise caution in our use of metaphor since, like the weather, it cannot be wholly predicted how it will behave. We must, if we are to address the moral failure Thorsby describes, be sensitive to the sources upon which we draw when crafting our metaphors and mindful of the social values invoked that may lead our metaphor astray.

So what metaphors to use? As Springer notes, environmentalists resisting the instrumentalization of nature frequently employ metaphors of agency and animism when doing so. Tim Christion argues that at least the metaphor of agency is going to fall short of the political goals it was employed to achieve. The value of the agency of nature metaphor is that it foregrounds the backgrounded ways in which the environment is active in restricting human agency, enabling us to live and to act in increasingly restricted ways. Since agency is thought to be the property of a subject on the dualistic, modernist understanding of humanity and nature, the idea is that affording nature agency will be a positive step toward decentring the human in order to create respect for the well-being of natural systems. In Christion Myers's view, the agency concept draws upon a problematic activity/passivity divide to which the phenomenological conception of experience gives the lie. Intentional experience is equally active (as a mode of engagement) and passive (things act upon the perceiver), and hence the metaphor of agency, which recalls the free powers of a subject, fails to grasp the situation of embodied beings. If we are to respect the phenomena, in both the ethical and Aristotelian sense, we must adopt what he calls an 'ethos of empowerment': interacting with the world as responsive to our presence, but also engaging with it responsively as opposed to through the unthinking play of habit. In this way, we can employ Carolyn Merchant's metaphor of partnership in a slightly different way, focusing less on the autonomous agency of nature and more on the form of relationship that obtains between humanity and its environs.

Irene Klaver's chapter takes up a different metaphor, meandering, as a way of imagining a different environmental future. Since the modern world has been characterized by an emphasis on efficiency, linearity and goal directedness, meandering has fallen out of favour because it seemingly embodies the opposite of these traits. As we become increasingly aware of complexity and the relationship of complexity to background conditions, however, there may be a renewed role for meandering: Exploring curved, looping paths rather than following a straight road, we see an alternative notion of progress and what it means to be efficient. While these claims may not seem intuitive, as Springer might note they are not supposed to be since meandering is a metaphor for the employment of a certain form of ecological imagination. Like any metaphor, then, it moves in two directions, here drawing upon the fact that the Meander is *actually* a river in Turkey. On the one hand, rivers carry sediment, laying beds through which they wend. On the other, this movement is achieved by the water's continual carrying away and displacing the sediment. Hence, meandering is a two-sided project of laying foundations, that is, making a start, and subsequently changing direction, reevaluating and beginning anew. To meander, then, is to embody a '*metis*', a set of habits of doing and thinking oriented toward overcoming difficulties through innovation and creativity rather than rule following.

After so much discussion of the importance of phenomenology to various issues of environmental concern, the final division of the book begins to show how the ideas and methods of phenomenology can be applied to specific areas of environmental concern, education, decolonization and aesthetics. While far from an exhaustive list of possible contributions—there is a rich phenomenological tradition of discussing technology, for example—the hope is that readers will see what Ingrid Leman Stefanovic, Zachary Shefman and Kristina Welch describe as the 'iterative' quality of the phenomenological approach within the world of practices: it is not the case that the application of phenomenology leaves the approach itself unchanged. As James argued above, a good phenomenologist will adapt and change their approach to be better suited to the object of inquiry. In this way Klaver's chapter is a fitting transition in that it closes with a detailed reexamination of the contemporary attempts to have the Los Angeles River 're-meander' from its current channelized condition. This last set of chapters embody that approach, displaying a concern for how phenomenology can enrich our understanding of environmental situations and must broaden its traditional concerns in order to do so. Here we see phenomenology at its most vital.

Stefanovic, Shefman and Welch open the discussion with an examination of the ways in which phenomenology has already expanded beyond the limits of philosophical inquiry and continues to inform the interdisciplinary approach to environmental studies. Indeed, the authors survey an array of ways in which phenomenology has influenced a variety of interdisciplinary fields, arguing for the centrality of phenomenological concerns when facing so-called wicked problems and examining the emergence of novel forms of interdisciplinarity as a result. The centrepiece of the chapter is a case study of how environmental studies programs evolved within Canadian universities. While a somewhat specialized topic, the authors open up a discussion of the role of theory in general within environmental studies and demonstrate convincingly the reciprocal relationship between theory and practice if we are to address the environmental challenges facing us today.

Barbara Muraca attempts just such an engagement with respect to the concept of ecosystem services. The notion is usually appropriated by economic-centred development discourses, but Muraca makes a case for why the concept itself, when embraced within a nondualistic conception of nature and humanity, can possess libratory power for indigenous peoples. While the dominant use of the ecosystem services concept allows policy makers to instrumentalize the environment, reducing people's living relationship to their environs to a monetizable commodity, the concept has the advantage of framing environmental concerns in terms of relationships. In Muraca's view, as has been a theme in the volume to this point, one of the main advantages of the phenomenological approach is to challenge the oppositional definition of humanity and nature. This rethinking brings to the fore an alternative way of approaching the concept of ecosystem services: Rather than have humanity retain its sovereignty over a nature regarded as only an instrument for human use, we begin to see how humanity is transformed through the use of nature and dependent upon it. At the same time, understanding what services an ecosystem requires a heightened attentiveness to the specificity of the relations that constitute that system. Such attentiveness creates space for the importance of local knowledges and, hence, the empowerment of indigenous peoples to articulate the particular eudaimonistic relations to the environment. In this way, thinking through the ecosystem services concept can be a promising approach for environmental justice.

But how are we to understand the challenges indigenous peoples face with respect to the environment? As Kyle Powys Whyte explains, the struggle for environmental justice does not *only* pertain to the way in which power is structured through institutions that disempower groups but *also* needs to include an understanding of the way environmental degradation contributes to the impoverishment of indigenous peoples' lifeworld. This impoverishment takes the specific form of taking from a group the ability to experience the world as a place capable of supporting their communal form of life. We are all embedded within a particular social-ecological context that is characterized by cultural and power laden relations between humans, nonhumans and ecosystems. From this context emerges a system of responsibilities oriented toward preserving that social-ecological context. As various forms of pollution erode the environments that sustain the practices of a people, the system of responsibilities is also washed away resulting in the loss of meaning and purpose. Of course, social-ecological contexts and thus systems of responsibility change over time, and so Whyte defines the injustice in terms of a concept he calls 'collective continuance': An injustice occurs when one group introduces changes so rapidly that the other group is left without the means to adapt to said changes. The collective continuance of one group is assured at the expense of the collective continuance of another's.

While Whyte is concerned with the destruction of the lifeworld, David E. Cooper's chapter examines the way in which our sonorous existence can transform our relationship to it by fostering a sense of ontological unity with the rest of the natural world. Music, in his view, makes nature more present to us in a variety of ways, especially in music's almost cosmological function in expressing a particular relationship to self, others and environs. Music is a craft that brings together natural materials and human embodied gestures to produce something more than would have existed were the two to have remained separate. Moreover, even listening to music is an embodied process of engaging with the natural world: The rhythmic connections to movement call upon us to move and feel along with the music. These bodily involvements call attention to the intimate ties that bind us to the natural world, inspiring numerous composers both to utilize different aspects of nature as a model and to compose for nature. The bodily experience of our sonorous connection to the world in music reveals the codependence of nature and humanity. When Cooper speaks of codependence, he has in mind something akin to the Buddhist notion of interdependent co-arising or the Nietzschean-Deleuzian notion of becoming: the interplay of differences that mutually shapes the development of each. In this way, Cooper's findings here with respect to nature hearken back to the metaphysical claims of the first set of chapters, revealing the relational constitution of nature and humanity in the manner in which music culturally informs the experience of nature and the experience of nature informs the formation of that culture.

Cooper's insights into music are then broadened out in the volume's final chapter, 'Phenomenological Aesthetics of Landscape and Beauty'. Here, Guðbjörg Rannveig Jóhannesdóttir develops a phenomenological conception of natural beauty with the goal of enabling aesthetic concerns to be integrated into environmental decision making. Like Cooper, the entry into this alternative conception of beauty lies in our bodily immersion in a landscape. This notion of the relationally constituted self leads Jóhannesdóttir to the conviction for which I argue, namely that meaning and value are inherent in the body's relations to its surroundings. This being the case, beauty is not a property possessed by the environment, but a specific quality of relation between perceiver and perceived. Specifically, the experience of beauty is characterized by a rapt attentiveness coloured by desire, that is, an opening up to appreciating and being affected by that which is being perceived. This type of experience is one that decentres the self: We are drawn out of our everyday modes of relation in the attempt to sustain the experience by engaging with the beautiful. Because we are engaged, drawn out of ourselves in appreciation of the landscape and attentive to the ways in which we are shaped by the landscape, the experience of beauty can be helpful in the policy context: The experience of relationality highlights the way things are (and are not) related, the layers of narrative present in the landscape and different ways of engaging with the landscape that allow a positive human presence within the landscape.

Seemingly disparate in content, these chapters cover a broad array of topics and terrain. Again, they are not meant to present any one conception of or approach to phenomenology, but rather to demonstrate a variety of ways in which phenomenological thinkers engage with environmental issues. That being said, the reader may already see the emergence of certain commonalities between the authors. For example, in beginning with experience phenomenologists tend to engage in descriptive analysis that leads us not to distinguish so-called subjective feelings from the objects to which they relate. The result is the important acknowledgement that our human thoughts and feelings *matter* when engaging with the natural world. They are, as a part of nature's fabric, part and parcel of effective responses to environmental problems. We can no longer hold ourselves apart and argue *only* for the preservation of wild spaces far from our human places. The chapters here speak of the need to integrate preservation practices, as Steven Vogel (2015) would have it, into the world that environs us, including the built one. They also speak to the need for an increased attention to the connections between social problems and environmental decay. Rather than disengage from nature, we need to engage with it differently.

Another commonality here is the importance of interpretation, of perspectivism and expression. Nature is the background on which all meaning is possible, we draw the substance of our lives from it, but can do so only as a result of relations between an affected and an affecting body. The authors here take this co-constituting relation of meaning and background as fundamental to our experience, which means that all knowing is knowing from a position and is therefore subject to interpretation and reinterpretation given changes in the conditions of experience. For this reason, and as several authors note directly, environmental philosophy is an inherently political project for a phenomenologist, one embroiled in the type of narratives we utilize in order to make sense of the world. While the emphasis here has been on reinterpreting concepts and rethinking methods and approaches, there is also significant reflection on the nature of specific practices, habits and values, all of which take place in the social world on the background of nature. Doing so is not merely an application of method to a case, but a way of deepening and exploring the implicitly political project of phenomenology: making explicit the narratives that emerge from the lifeworld, exposing the habits and practices that structure those narratives and subjecting them to analysis and critique.

Following from that point, I will lastly note the broad agreement that the focus of philosophical analysis pertaining to nature ought to be *relations*. Phenomenology is a relational philosophy and is therefore amenable to allying itself with forms of political and social philosophy such as feminism or critical race theory that also seek to understand the political world through relations of power. In the environmental context, such analyses can help us to better understand the nature of social-ecological systems and therefore what may be more or less productive ways to engage in those systems. This point is of additional importance given contemporary discussions of the anthropocene, the age in which human activities are taken to be the dominant driver of environmental change. Focusing on relations shifts the normative question in the direction of inquiring into how we might inhabit nature more justly, decently and more welcomingly to the other species with which we share the planet. In short, how might we inhabit the Earth in a manner befitting the values we claim to espouse?

Part I

Phenomenology of Nature

ONE

Mythic Enlightenment

Phenomenology and the Question Concerning Nature

Bryan Smyth

This chapter is concerned with a certain contribution that transcendental phenomenology can make to contemporary environmental philosophy, and to environmental ethics in particular, with regard to the task of articulating a critical hermeneutic alternative to the traditional and still-dominant conception of nature. Although my discussion will have implications concerning the content of this hermeneutic view, the focus of my discussion will be on its idiomatic form. Drawing mainly on phenomenological considerations, the thesis for which I shall argue is that nature is best conceived in the idiom of *myth*, understood as *a dynamic narrative horizon of significance*.

This thesis makes an epistemic claim, and it is opposed primarily to any logos of nature, that is, to any determinate ontological conception of nature held as objectively true and knowable. The point is that myth (or mythos) so construed offers a positive way forward with regard to 'the question concerning nature' by enabling a viable middle course to be steered between, on the one hand, unwitting acquiescence in the problematic dualisms pertaining to the dominant conception of nature, and, on the other hand, the wholesale dismissal of the notion of nature that worries about such acquiescence can motivate. Myth can do this because, unlike logos, as a narrative horizon of significance it can posit nature without the reification of it (i.e., its objectification as a *thing* with invariable properties) that is a necessary and (for all intents and purposes) sufficient condition of those dualisms. It thus enables us to retain a non-anthropocentric conception of nature as the framework for critical environmental reflection, yet without saddling that reflection with the aporetic task of identifying non-anthropogenic sources of normativity.

This approach will thus have a direct bearing on questions concerning ethical motivation with regard to environmental issues, and on this score it can be given a fairly straightforward and plausible justification. Still, in theoretical terms, any talk of myth can raise alarm bells as an irrational or reactionary regress wholly antithetical to critical thought. Against this, my key claim will be that a turn to myth is theoretically justified because it is *rationally inescapable* when one must reckon with the outer limits of experience. This is where phenomenological reflection kicks in, for it shows that human experience in general is situated within a horizon of meaning that in principle cannot be apprehended objectively, but which operates solely as the ultimate background for any such apprehension. This horizon may (and typically does) differ across different individuals and sociocultural contexts of experience. But in each case it amounts to the referent of the idea of nature defined stipulatively as the inclusive totality of reality. In this way, nature is an ineluctable (though varying) feature of human experience in general. It *does* exist, so critical environmental thought must come to terms with it. And this can only be done, I submit, in mythic—and moreover, as I shall further contend, *mythopoetic*—terms. This approach is consistent, even to the point of remaining within the broad ambit of humanism, with the rejection of any *concept* of nature, or the 'denaturalization' of environmental thought (Vogel 2002, 2015; Biro 2005). It's just that in retaining a *myth* of nature we can more successfully articulate a humanism that has a non-anthropocentric normative profile because we can now subscribe to a form of humanism that just *is* non-anthropocentric in virtue of being based on an unreified conception of nature.

Although I can't explore it in detail below, the larger claim standing behind my argument is that mythos is a primordial but also permanent, essential and irreducible dimension of human consciousness, not an archaic stage with which logos effects a decisive break in phylo- and ontogenetic terms.[1] The sort of rehabilitation of myth that I have in mind is therefore the key to overcoming the objectionable dualisms of traditional thought (e.g., culture vs. nature, human vs. animal, mind vs. body) because the archetypical dualism therein is precisely that between logos and mythos. In shifting its view of nature within *this* register, critical environmental thought can reengage holistically the ontological continuity believed to have priority over those binaries and on that basis generate a relational standpoint adequate to the normative complexities of the contemporary world. This mythic perspective has ramifications for a wide range of fields. But it is entirely fitting that it be worked out first in the context of environmental thought. For it is here, precisely in the form of 'the question concerning nature', that the problem of totality imposes itself most directly, and thus in a way that pushes the envelope of enlightened thought most radically.

The discussion will unfold across four sections. (1) I will first rehearse the chief motivations for pursuing a new hermeneutic conception of nature, and discuss the practical considerations that Baird Callicott has proposed in favour of a mythic view. (2) In pursuit of a theoretically stronger argument, I shall then discuss the phenomenological notion of 'horizon', and argue that nature is best conceived as the ultimate or outermost horizon of experience. (3) Informed by Anthony Steinbock's work, I will then contend that, so understood, nature is, in phenomenological terms, a *generative* notion, and that as such it implies the normatively charged methodological task of projecting the outer horizon of experience—that is, articulating what the meaning of nature is or what the totality of reality *should* be.[2] (4) Bringing this together, and drawing on Hans Blumenberg, I will then argue that nature qua horizon is best conceived in mythic terms, that is, as the projected result of a historically dynamic mythopoetic process of significance-bestowing narrative.

TOWARD A MYTHIC HERMENEUTIC OF NATURE

Several reasons may be adduced as to why it is necessary to raise 'the question concerning nature', that is, rethink the *meaning* of nature. Of these reasons, two are particularly salient. First, despite notable progress, traditional conceptions of nature that are complicit with the practices which have led to current environmental crises remain effectively hegemonic within the population at large. This limits the impact that critical environmental activism can have. For even if more people than ever are familiar with ecological science, in normative terms this all too often lacks effective motivational force. Betraying (among other things) the stubborn perdurance of traditional views of nature, this situation presents the root reason for critical environmental thought to want to rethink what is meant by nature.

The second reason is that there is no universally agreed-upon answer to 'the question concerning nature'. There is, rather, disconsensus among environmentalists with regard to the basic world views that inform their work—a situation that results in incompatible and even conflicting normative conclusions (e.g., concerning restoration). This is troubling, at least inasmuch as the larger goal is to cultivate as widely as possible a new environmental ethos—if even dedicated activists are talking past one another, then it's hardly to be expected that broader strata of society will come on board.

To be sure, some of this disagreement is insurmountable—between some environmental views there is indeed no meaningful common ground. Still, it may reasonably be hoped that most do share a common core, and that this could be expressed as an answer to 'the question concerning nature' in a way that would support the formation of a more inclusive and unified social movement. For many, ecological science is the obvious candidate for this kind of environmental lingua franca. As intimated above, however, it is doubtful that increased scientific awareness, however indispensable it may be, can alone serve as an antidote to apathy, indifference or disenfranchisement in the ethical sphere.

Here we face a choice: either try to rethink nature, or else discard the idea of it altogether. Those opting for the latter prong do so primarily on the basis of arguing that the very idea of nature belongs to the dualistic edifice of traditional anthropocentrism and that it is therefore irredeemably compromised. Good intentions notwithstanding, any attempt on *our* part to think *our* way out will inevitably reinforce that edifice. Indeed, they would say that the very idea of an environmental ethic geared toward human treatment of nature understood as dualistically other cannot but reaffirm an exceptionalism that issues either in a warmed-over anthropocentric humanism or else, what is merely its dualistic flipside, a radical misanthropy. Nature is, in this view, a conceptual trap. It would therefore be best to do 'environmental philosophy without nature' (Vogel 2002, 23), or to 'de-naturalize' the project (Biro 2005) in the sense of expunging the idea of nature, in order to break free of the conceptual strictures it implies and to be able to articulate defensible normative conclusions on a coherent conceptual basis, even if it means that that basis remains one of humanism.

As noted above, the approach that I wish to outline in this chapter resonates sympathetically with this view. Yet it seems to me that the latter shares with the approaches it purports to reject the injudicious and unnecessary assumption that what is at issue is the *logos* of nature. With the rejection of any such approach as bearing the germ of anthropocentric domination, I entirely agree. For it involves a reification of nature which, even if it is itself normatively neutral, entails the ontological discontinuity that subtends dualism and the ethical problems thereby implied.

But there is a third option, viz., rethinking nature, but doing so in the idiom of mythos.[3] The idea is that such an idiomatic shift, because it aims to articulate significance without advancing a claim to ontological truth, can support an unreified conception of nature. The value of this has to do with the problems noted above. In particular, it has the potential to provide critical environmental thought with a relatively uncontroversial basis for internal consensus. For if concerning nature there are deeper core intuitions common to otherwise theoretically discordant views, then these might be expressible in mythic terms without foreclosing debate on other issues. More broadly, expressing an environmental ethos in mythic terms could help with effectively broadcasting the message not only in cognitive terms but also at an affective level that could have greater motivational impact.

Baird Callicott has raised similar ideas in arguing for a mythic view of nature as the ground for environmental ethics (Callicott 2002, 2011). His basic argument runs as follows: (1) Important environmental issues are essentially global in nature and as such require a coherent global response. (2) Human beings live in a multiplicity of cultural contexts—to be practically effective, then, environmental ethics must be expressed in 'the grammars of local cultures' (2002, 158). (3) Environmental ethics therefore needs a framework that is 'universally intelligible and acceptable' (2002, 162), yet compatible with cultural diversity. (4) Science in its modern sense does *not* supply an appropriate framework. For in claiming unique access to truth and objectivity it involves an 'epistemic arrogance' that is incompatible with cultural diversity (2002, 162). At work here is the postmodern 'deconstructive' recognition that there can be no unmediated access to reality, and that representations of it are always socially constructed in and through discourse. (5) This recognition does not derail or invalidate science. But it does mean that it is always embedded hermeneutically within a narrative framework or mythic 'world view'. (6) Although such world views make no claim to their own truth, they condition what *counts* as true and real, and can be judged as more or less *tenable* according to several specifiable criteria: internal consistency, consistency with scientific results, comprehensiveness, revisability in the light of new experience, 'aesthetic and spiritual appeal,' and, crucially, the extent to which it 'facilitate[s] the survival and prosperity of its subscribers' (2002, 167ff; cf. 2011, 516f). (7) Alongside a deconstructed view of (ecological) science, then, what is needed is a '*re*constructive' approach that takes up, explicitly and self-consciously, the open-ended task of formulating a mythic view of nature as tenably as possible in the context of contemporary critical environmental thought. This myth would be a new 'grand narrative' that reaffirms the category of totality. It's just that the idiomatic shift to myth means that questions concerning practical tenability are fully distinct from those concerning theoretical veracity (thus dispelling worries about illicit totalization). Callicott's hope is that by pairing science with myth along these lines, 'reconstructive post-modernism' would be able to provide an optimal grounding for environmental ethics by successfully bridging universality and difference through the ascription of an appropriate normative significance to the results of science.

In terms of content, the mythic narrative that Callicott has in mind relates closely to the Leopoldian land ethic of which he is, of course, a leading advocate (Callicott 1989). It amounts to what he calls 'the evolutionary-ecological myth' (2002, 168) or 'worldview' (2011) which, although informed by science, goes beyond it in assigning axiological significance to evolutionary history and ecological holism in such a way as to bring it about, for example, that 'other species are perceived as our phylogenetic kin and working partners on the economy of nature' (2002, 162) or, more generally, that we might begin to 'think like a mountain' (Leopold 1949). The implied narrative satisfies fairly well the general criteria that Callicott laid out, including on the practical side. For in emphasizing themes of embeddedness, cooperation and wholeness, '[it may] inspire its subscribers to better adapt, long-term, to the ecological exigencies of the biosphere, and thus prolong human tenure on the planet' (2002, 169). The key point is that this ethical inspiration, and the perceptions upon which it is based, do not derive from evolutionary or ecological science directly, but from the meaning attached to their results by the adopted mythical-narrative framework (2011, 521).

More could be said about Callicott's position on myth. But for present purposes it suffices to have shown that a turn to myth in environmental thought is cogent and well motivated, and not a whacky or madcap proposal. Still, objections may be raised against it. In moving forward, then, I just want to raise one point of criticism, to wit, that in his discussion Callicott offers a mixed rationale. The turn to myth is presented as following primarily from the inability of science to provide normative grounds for environmental ethics. However, Callicott also suggests unmistakably (e.g., 2002, 158, 171; 2011, 517, 520) that the turn to myth is based on the idea that science is *too difficult* for 'ordinary folk' to understand, and that its ethical uptake therefore requires the message to be communicated in a simpler form. This claim plays into standard objections to myth that see it as ineluctably compromising rationality in some way. As a qualified defence of what I take to be Callicott's main claim, then, I want to show that far from any potentially disputable practical considerations, there is *an essential theoretical need*, based in the very nature of the experience of nature, for critical environmental thought to make a mythic turn.

PHENOMENOLOGY OF HORIZONS

To this end, I shall briefly discuss the phenomenological notion of 'horizon'. The upshot here will be that nature is best conceived as the ultimate or outermost horizon of experience, leading to the claim that it falls within the domain of myth.

The phenomenological notion of horizon is rooted in, but not identical to, the ordinary pre-philosophical meaning of the term. Consider, for example, the paradigm case of the horizon one experiences when looking out over the ocean, or a wide-open landscape. Three interrelated aspects of this experience are salient (cf. Kuhn 1940, 107–8; Geniusas 2012, 1–2). First, the horizon is experienced as a circumscriptive limit to one's experience—yet it is also something which, in *framing* that experience, serves to determine its content. Things appear in meaningful ways *within* the horizon at least partly *in virtue of* the horizon, which itself remains, in a sense, invisible. Second, we do not experience the horizon itself directly, but only indirectly in shifting our attention away from particular objects. In doing this, however, the horizon always recedes further—it is an essentially open liminal structure of our experience which we can never, in principle, surpass. Third, as a structure of experience, the horizon is essentially relative to one's subjective situation—even though it conditions one's experience of all things, the horizon is not straightforwardly *real* in any objective sense. Otherwise put, there is no such *thing* as the horizon.

So considered, the ordinary notion of horizon parallels closely the Gestalt-theoretic notion of 'background', in particular with regard to how the structure of conscious experience is in general a matter of a focal object, of which one is explicitly or thematically aware, appearing against a background, of which one is only tacitly or apperceptively aware, yet which conditions the focal experience.

Phenomenology can be understood as the general attempt to ground philosophical analysis on a metaphysically unbiased descriptive account of *what really goes on* in experience. For this reason, the notion of horizon is absolutely central to it. Phenomenological description is concerned with the intentionality of experience, and in the context of the epoché and reduction, the rigorous pursuit of such description leads inexorably to the exploration and elucidation of the apperceived horizonality of experience, and ultimately to consideration of what may be referred to as 'the horizon of horizons'.

It goes basically like this. Consider the given contents of your experience precisely as they are meant or intended, that is, without any metaphysical presuppositions as to what they 'really' are. What quickly comes to light is that the intentional experience of anything whatsoever always exceeds what is actually given directly in that experience. The standard elementary example is of a cube—although we can never see more than three sides at once, the experience of it is always (at least in normal cases) of a six-sided object. And this applies to intentional objects in general—what is given directly is always given within a sort of tacitly apperceived 'halo' of potentialities, the gradual unfolding of the ensemble of which leads to a more intensively complete apprehension of the intrinsic nature of the object in question. Such, in a nutshell, is what Husserl termed the 'inner horizon' of an object. The object itself is experienced most fully and concretely through the synthetic aggregation of successive moments of awareness that serve to unpack its tacit horizonal dimensions. Leaving aside the question of whether and how my experience of the object lines up with yours, let us just note that this movement of increasing awareness always involves some dimension of past recollection and futural anticipation, and that the inner horizon of an object is ultimately therefore a temporal or historical horizon.

Even this remains abstract, however. For objects are never experienced in isolation. All things are related to other things, and they are situated within different contexts and environments. And these connections are not incidental to the intentional reality of the object, but are tied to it through relations of mutual implication which, again, form the tacitly apperceived background of experience. Such is the gist of what Husserl termed the 'outer horizons' of an object. In analogy with the inner horizon, the progressive elucidation of outer horizonality leads to a more extensively concrete and complete apprehension of the object. And it likewise has a temporal or historical structure, and this intertwines with that of the inner horizon. This is the sense of the sort of unbiased concrete experiential evidence that phenomenology seeks as the grounding for philosophical analysis.

Horizonality in general, but especially on the 'outer' side, can be approached as a series of nested levels, and in each case what is originally present in experience as background *can*, through deliberate effort, be foregrounded and thematized. It's just that *that* is then experienced against a further level that itself necessarily remains tacit as background. If this process were to continue indefinitely, then phenomenology would face a vicious regress. For the aim is to gain a critical and concrete apprehension of the contents of experience through the active elucidation of its horizonal intentionality. If phenomenology could not carry out this elucidation completely, then its results would be uncritically provisional, and its philosophical pretensions would be subverted. On the assumption, then, that phenomenology can indeed make bona fide philosophical claims, there must be an ultimate 'horizon of horizons' that is accessible to it. This 'horizon of horizons' or outermost horizon is best conceived, I submit, as nature understood as the inclusive totality of reality. This view challenges the assumption that phenomenology properly begins at the level of discrete objects and then proceeds upward, so to speak, to broader and more inclusive views. Such an assumption is widely held, but it is mistaken. Careful meta-level reflection shows that the logical order of priority is the reverse: The concreteness of phenomenological description correlates to the degree of horizonal elucidation such that methodological primacy ultimately attaches to the elucidation of the outermost horizon of experience (i.e., nature). The idea is that if we're not clear about this, then in principle we can't have a critical view of anything at all, since everything falls within this horizon and is, at least to some extent, conditioned by it.

PHENOMENOLOGY AS GENERATIVE

This view seems, however, to present some serious difficulties. In particular, if experience in general is structured horizonally, then even if inter-mediate-level horizons can be brought to givenness and thematized on that basis, in principle an ultimate horizon could *never* be given, since to be given *means* to be given within a further horizon. It would seem, therefore, that the horizon of horizons can never be anything but a horizon, and that therefore nature understood in that way can never be accessible phenomenologically.

This objection is well taken but it misses the mark. I would agree that nature as such cannot be *given* in human experience. But it does not follow from this that nature is radically inaccessible. The basic rationale for conceiving nature as horizon is to avoid the dichotomy between immanence and transcendence by locating it precisely *between* these dimensions. This is how it can be that, with regard to human experience, it has an omnipresence that is neither fully within nor without.

Yet the question remains: How exactly does phenomenology relate to nature as this kind of horizon? The short answer to this question is that phenomenology is ultimately a *generative* project (Steinbock 1995, 2003). Although this is not obvious to begin with—and may even seem, at least initially, to conflict with the guiding motivations of the project—phenomenology *cannot* simply be a descriptive project. For even if one worked out a comprehensive genetic account of one's experience on the basis of its apperceived horizonality in all its aspects, one could not be satisfied because one would have to admit that even that account remains conditioned uncritically by whatever further horizons within which it is located. It is a transcendental condition of human consciousness that there be an ultimate horizon. Typically this horizon is established, revised and operated within experience in a passive, unself-conscious and hence completely uncritical way. And this applies no less to the most rigorous moments of phenomenological description. Here is an arbitrariness that cannot but come to be seen as theoretically and—especially—normatively unacceptable. Yet the basic phenomenological approach itself is too valuable to jettison.

What phenomenology can do, however, is provide its own foundation by recognizing ultimate horizonality as a transcendental feature of conscious experience *and by addressing it in an active, self-conscious and critical way*. This is the definitive insight of generative phenomenology. In working back through static and genetic layers of intentional analysis, phenomenology exposes—contrary, perhaps, to its original intentions—the naïveté of pure presuppositionless description. For logically prior to the project of description is the generative *projection* of the outermost horizonality within which the contents of experience that are to be described in terms of their intentionality first appear. This projection amounts effectively to stipulating what the meaning of nature *should* be (cf. Steinbock 1998, 189f, 196), as opposed to the received meanings that it might have in this or that historical or cultural context, and this moment of irreducibly normative generativity can be regarded as a fundamental gesture of the phenomenological reduction. In this sense, then, nature is a generative notion for phenomenology—indeed, the most basic one—and phenomenology's ultimately issuing in the hermeneutic task of rethinking it as defensibly and responsibly (or 'tenably') as possible is, I think, entirely salutary for critical environmental thought.[4]

NATURE AS MYTH

My principal claim—viz., that nature is formally best conceived in the idiom of *myth*, understood as *a dynamic narrative horizon of significance*—can be seen to emerge fairly directly from the preceding considerations. Just a few points of clarification are necessary in order to show that the outermost horizon of experience is best understood in those terms that are being used here to define myth.

Concerning 'significance'. The ultimate horizon of experience—that is, the conception of nature that is tacitly operative in one's experience—does not determine any particular experiential content. Rather, it operates as the outermost background that contributes in a crucial way to the *meaning* that any such content will have. In this way there is no question or issue concerning its being true or false. Nor, then, is there any question of whether it is *believed* to be true. That is all beside the point. It is simply a matter of the fact that different tacitly held conceptions of nature will lead to different experiences of objects which, considered more abstractly, appear the same. This imposes the need to judge which experience is *better*. In analogy with the effort one may make to regard an artwork, say, or a historical event, in the best and most appropriate way—that is, in a way that shows it most accurately and revealingly—by shifting one's point of view *in order thereby to alter the relevant background horizons*, the need to engage generatively at the level of the ultimate horizon is driven by a concern with what is in the foreground and experiencing this as well as possible.

This is obviously a normative rather than a theoretical question. More specifically, using the term in the sense given to it by Blumenberg, ultimate horizons are matters of '*significance*', and the function of myth is to provide such horizons. Significance is here understood as a precognitive framing or orientating of the totality of experience that endows or 'charges' it with a normative or 'valenced' sense of meaning and purpose (1985, 67f). Following Gehlen (1988), Blumenberg contends that, unlike other species, humans are 'always not yet determined' in the sense that they do not relate to their environment in a strongly determinate way and can alter it to a much more considerable extent. Confronted initially with an indifferent and overwhelming world, humans have an existential need for significance as a kind of transcendental condition of disalienation or feeling at home in the world (cf. Rose 2007, 637). Needless to say, of course, the mythic projection of significance has, especially in the modern world, typically been enacted in highly problematic ways—disalienation has in vain been sought through domination. This is not, however, because significance was projected, for that's a transcendental necessity. Rather, it is because the projected significance was normatively deficient. And this is because it was typically underwritten by some form of logos, a theoretical discourse claiming ontological truth. This is the problem. For significance is logically prior to questions of truth. What is crucial to recognize is that myth is not a primitive form of logos that is oriented teleologically toward and superseded by modern scientific rationality. In structural and functional terms, myth differs qualitatively from logos, and the same applies to outer horizonality vis-à-vis the contents of experience—myth provides the horizon of significance against the background of which a scientific logos of those contents can be developed. In no way, therefore, does engaging in the projection of horizonal significance compromise one's commitment to scientific reason. Nor need it commit one to anthropocentrism. To be sure, there are certainly some such risks involved. But the view I am proposing does not introduce these risks. Rather, it just brings them into the open by shifting the primary burden of rationality from the objective discovery of nature to the acknowledgement and acceptance of normative responsibility for the *meaning* of such discovery.

Let me just add two further points. First, concerning 'narrative'. The significance represented by nature qua ultimate horizon is best understood in narrative terms. In general, narrative is a sequence of events, the structural coherence of which assigns to each a certain meaning depending on its position within the whole. An experience is thus made sense of by inserting and locating it within an overall narrative structure. In particular, just as static phenomenological accounts of something are augmented by being set within a genetic account of its historical unfolding, that genetic account is itself augmented by being set within open-ended temporal horizons that impose a broader developmental logic—an overall sense of whence and whither. It is especially in this regard that a generative conception of nature is irreducibly normative—it is by projecting an overall historical significance that one effectively stipulates what the meaning of nature should be, and this in an intentionally self-fulfilling sort of way. To be clear, I am not arguing for the *introduction* of a mythic conception of nature into environmental discourse which would otherwise be free of myth, but rather for the introduction of a *new* myth of nature, the underlying point being that as a matter of transcendental necessity experience is always framed by *some* mythic narrative, and that the critical task is to affirm remediatively a more acceptable (or 'tenable') mythic *counter-narrative* to existing hegemonic ones (cf. Plumwood 2005). If there is a self-fulfilling tendency among the transcendental conditions of our experience, then it is incumbent upon us to ensure that it is oriented in the most normatively defensible way.

Second, concerning 'dynamic'. Myth is a narrative that supplies experience with the outer horizon of significance, but it does this across changing circumstances. Contrary to some common assumptions about it, myth is not necessarily set out once and for all, but rather is something that is, at least potentially, in a constant state of evolution. That is to say, myth is a process that consists essentially in what Blumenberg called '*work on myth*' (1985), the ongoing transformation and revision of the framework of overall experiential significance. This mythopoetic work, which is inescapable, typically occurs more or less passively and unselfconsciously. But it can be expressly motivated in various ways. In particular, as in the case of environmental thought, it can be motivated by the recognition that prevailing and predominant myths are normatively deficient and problematic. Yet although it is primarily a normative motivation, working out mythopoetically a better experiential horizon does need to take cognizance of the results of scientific reason. There is a dialectical codependency between logos and mythos that needs to be affirmed. Myth should be consistent with science, and so must follow it; yet on the other hand, science itself remains dependent on myth. This is the most important point: although essentially bound up with it, myth is irreducible to logos, and in principle its task cannot be performed by theoretical discourse. When the latter disclaims myth and implicitly usurps its function, however well intentioned it may be, the possibility of taking responsibility for the horizonal significance of experience is precluded, and that's when serious problems can arise.

CONCLUSION

The foregoing considerations support the claim that nature, understood as the inclusive totality of reality, is best understood in phenomenological terms as the outer horizon of experience, and that as such it is best conceived as a dynamic process of significance-bestowing narrative—in other words, as myth.[5] Although I have not tarried over the distinction, it should be clear that the view I am proposing has nothing to do with the usual pejorative connotations of myth as being untrue and unreal. The key claim is that a mythic outer horizon is an essential and inescapable aspect of human experience, and that it is therefore a theoretical strength to be self-consciously aware of this fact, and to reckon with it generatively in the most rationally and normatively defensible way. We need not—and should not—simply resign ourselves naïvely to whatever myth is passively given, but as a matter of rational and ethical necessity we need to intervene actively and take responsibility for the horizonality of our experience. To be sure, the mythopoetic task with regard to nature that generative phenomenology implies is, as noted above, fraught with certain risks, inasmuch as it effectively politicizes the question concerning nature and makes it the site of what amounts to an ideological struggle. But at least in the environmental context, where it can be asserted with confidence that the prevailing horizon of experience is grievously deficient, the perils of blind acquiescence far exceed the risks of mythopoesis. The broader underlying point is that far from being some sort of misological or otherwise flaky reenchantment of nature, this sort of mythic turn represents—counterintuitively perhaps—the pinnacle of enlightened reason. While this may downgrade the latter's theoretical pretentions, in opening up issues concerning the outer horizon of experience and thus how reality should be encountered, it raises the bar of intellectual ethical responsibility significantly higher.

NOTES

1. Recent criticism of the traditional view of the relation between mythos and logos includes Stambovsky 2004; Bottici 2007, 17–43; 2008; Rose 2007; Tofighian 2010; see also Coupe 1997.

2. Although not a standard view in phenomenology, this perspective is clearly supported by the basic aims of the project—for an opposed view, see Crowell (1996). The matter ultimately hinges on the methodological question as to whether the outer horizonality of experience can be addressed critically.

3. My view has some general points in common with Scott Cameron's reply to Vogel (Cameron 2014).

4. This approach is broadly congruent with the project in Bannon (2014).

5. Here we may note a pair of related advantages that this view provides. First, it is consistent with the idea that nature as such is always somehow beyond or in excess of human experience—it's just that this excess is not a matter of radical alterity but of the horizonal character of nature and its priority within human experience (cf. Foltz 2004). Second, this view also resolves the vexing equivocation between 'Nature' and nature (nature-in-itself vs. nature-for-us). I have eschewed any such contrastive talk because in the unreified conception that the mythic perspective affords, the referents of this distinction coincide unproblematically.

TWO

Towards a Phenomenology of Nature

Janet Donohoe

Phenomenology has traditionally been understood to focus on universal structures of consciousness that make experience possible. Many suggest that this makes nature merely a correlate of consciousness, thus indicating that nature could not possibly be universal or structural. Others argue that this understanding of phenomenology embraces an anthropocentric viewpoint that undermines any intrinsic value of nature.[1] While these positions may be true on a particular reading of phenomenology, I would like to suggest here that a genetic phenomenological account of nature reveals that there is nothing natural about nature, and that in spite of nature's cultural embeddedness, there can be universal elements of our experiences of it.[2] These universal elements can only be discovered through a genetic phenomenological account of experience of nature. Genetic phenomenology is characterized as an asking back into the sedimented layers of our everyday approach to the world, our natural attitude.

Sedimentation is the layering of meaning that accumulates through our repeated experiences over time. Some layers of sedimentation are inherited from previous generations as when our parents teach us that hot stoves are dangerous objects. We learn to approach hot stoves as things around which we must be careful. Likewise, our further experience of stoves as hot adds layers of sense and depth to our experience of them. Our subsequent learning to cook on a stove adds even more depth to our sense of stoves as we come to understand their role in providing food and as a focal point for the joy of cooking or the frustration of not being a good cook. The stove is culturally meaningful for someone in Western culture in a way that it might not be for someone who has grown up in a culture where cooking is done over an open fire. Genetic phenomenology, then, allows us to peel away layers of cultural sedimentation that characterize our everyday sense-making of everyday experiences in the natural attitude. Peeling away layers of sedimentation reveals that facts of the natural attitude are already laden with meanings that we have inherited from previous generations and from our own prior experiences. We never encounter things in the world in isolation or as mere objects. We encounter them as meaningful things even if we might not know exactly what they are. This means phenomenologically that we constitute things in our world, or we make sense of them, with multiple layers of meaning from previous experiences and from a base of culturally inherited ways of being in the world with things.

At the same time, we can recognize that there is a fundamental, universal level of experience of nature that supports those cultural meanings. Although I experience a different world than my grandparents, a contemporary person from a radically different culture, or even a nonhuman animal, for example, it is at some level the same world. This position does not reduce nature to cultural construct because ultimately it views nature not as a thing, but as a ground and horizon of experience itself. Such a claim allows us to understand nature not as something that resists or opposes us, but something that makes our experiences possible and that is shared among all creatures be they human, nonhuman animal, plant or even rocky substrate. It is a position that allows us to stress the interrelation and interdependence of earthlings.

Initially, in this chapter I will explore a distinction between homeworld and alienworld, or culturally diverse worlds, and the universal structure of lifeworld (nature). By understanding this structure, we can grasp how recognizing a shared nature makes possible a different approach to our world that is not anthropocentric and that allows us to question the identification of nature with a thing over against us that opposes us. Instead it will forge a path to an understanding of nature as common and as the foundation for a web of interrelations among beings of the world. I will investigate how lifeworld functions as horizon and ground of *all* experience, suggesting why this conception of lifeworld can be helpful when considering our general approach to the environment and our concerns for environmental policies. Finally, I suggest a way to rethink our relationship to and understanding of nature and nonhuman animals. To recognize the lifeworld as a common substrate of experience means that we understand that it is not a thing existing independently and objectively that can be analysed, mastered or controlled. Instead, our approach to nature can be one of renewal and critique. Renewal prepares us to recognize those ways of thinking about and interacting with nature that we have inherited from our own culture while critique pushes us to move beyond those inherited assumptions to an open questioning of the value of those ways of thinking in order to dig down to a deeper level of commonality, a deeper level of universal aspects of the experience of lifeworld. This makes possible the rethinking of the lifeworld-consciousness relationship.

HOMEWORLD/ALIENWORLD/LIFEWORLD

Any experience we have is characterized by a meaning-giving act and that which is given. These are two sides of the same experience and they are distinct, but not separate. One cannot be without the other. In the experience of the tree outside my window as I sit at my desk and write, the tree is given to me or perceived as tree. There is not a series of sensory perceptions that I must synthesize in order to perceive tree. Instead, I perceive tree as tree. Both tree and I are necessary for the experience and tree is necessary as tree, not as colour, shape, texture, etc. Tree is perceived in one fell swoop, so to speak, in its fullness as distinct from my act of perception, but not as separate from it. In other words, nature cannot be without subject, subject cannot be without nature. We can certainly focus upon the perceiving consciousness in this process, and if we do so, viewing consciousness as the key to understanding this experience, then the tree and its world become secondary. Static phenomenology might explore this experience in such a way, and one could feasibly stop there. Genetic phenomenology, however, takes us in a slightly different direction as a supplement to the static phenomenological approach. It begins with the idea that the natural world is pregiven meaning that it is the presumed horizon of our experience of any object. While constitution is focused on particular objects that are given, such as the tree in the example above, the pregivenness of the natural world simply means that we have no need to point out the reality of the world to ourselves. The natural world is the very foundation of the constitutional process without which the tree itself cannot be given.

When constituting the tree, it is not as if the tree exists in empty space, without relation to anything else, as pure product of my sense perception. I cannot constitute the tree separate from the world in which both it and I are found. Insofar as tree is constituted, it is so within the horizons of a world. That world is a world that is quite familiar to me and in which tree already has particular connotations, but also a particular history with me. The tree is constituted with all its richness of relations and meaning within a world of familiarity to me. That world is not itself an object that is accessible without all the particular things of the world that I encounter. World is accessible only as horizon of the things that we experience. In other words, world is not the collection of all the things we encounter, but is the presumed background that makes possible the experience of anything at all. Genetic phenomenology acknowledges that we are always already aware of the world before we consciously turn towards it in analysis or reflection.

Edmund Husserl has called that world with which we begin and from which subjectivity or consciousness and its experiences can never be separated, the lifeworld. It is the environing world, the surrounding world of our everyday sense that grounds any scientific world that we might want to call objective. As our surrounding world, the lifeworld is yet more than simply the setting or the stage for the action of our lives. Lifeworld surrounds in all spatial directions, but surrounds temporally as well in the sense that it is historical. Thus, lifeworld surrounds and is inseparable from temporal horizons of life. Does this mean that lifeworld is subjective? Phenomenology maintains that it is a false division to establish an objective world over against subjectivity. In other words, there are structures of world that are part of our presumptions about lifeworld and these are structures that are universal, but not ahistorical or objective. For Husserl, examples of these universal structures are the spatiotemporality of experience or the experience of bodies, but not bodies in the sense of physics. Husserl suggests that 'To live is always to live-in-certainty-of-the-world. Waking life is being awake to the world, being constantly and directly "conscious" of the world and of oneself as living in the world, actually experiencing and actually effecting the ontic certainty of the world' (Husserl 1970, 142–43). Such an understanding does not postulate the world as an object over against me. It understands world and me to be intimately interrelated. Nature is the framework for all experience.

Genetic phenomenology allows us to peel away the layers of sedimentation that compose our homeworld way of meaning-making in experience. It also allows us to peel away the layers of meaning-making that compose the lifeworld way and are historical. For instance, contemporary lifeworld is such that we presuppose certain privileges that we have as rational creatures and we categorize all other beings according to our understanding of ourselves as rational. This type of thinking and way of constituting our world is not necessarily one shared by prior or what we questionably call primitive peoples. This particular shared way of thinking is not a priori, but is historical and is part and parcel of our lifeworld. In peeling away layers of sedimentation through the genetic method, we can see that such a position is not strictly cultural; it is historical and so much a part of our lifeworld that it is very difficult to begin to understand it as nonessential. Genetic phenomenology in revealing these layers of presumed meaning makes it possible for us to suspend our commitments to these ways of thinking in order to be able to think differently. It also reveals to us, however, yet other layers that we take to be truly a priori and that are part and parcel of what make us earthlings such that any attempt to escape our earth-bound way of being is folly. This is an issue we will take up again momentarily. Let us first evaluate homeworld and alienworld.

Homeworld indicates that world which is most familiar to us and by which we measure every other cultural world we might encounter. Homeworld is not statically bounded. Its boundaries may shift and change, include larger and smaller, more expansive and less expansive conceptions of world. Homeworld might be very specifically my house, but it could also be my neighbourhood or the region of the country in which I live or even my national connection. Homeworld establishes norms for my experience. These are the ways in which I constitute the world until some experience in the world challenges what I perceive as normal by not fitting in with my expectations. In such cases I am frequently brought up short or surprised in my constitutions of the world. Homeworld establishes for us what we constitute as normal. This is not a value judgment, but simply a judgment about what we expect and how we usually constitute the world.

Alienworld, in opposition to homeworld, is that which is unfamiliar and frequently challenges those things that I take to be normal about the world. Homeworld and alienworld are not just places, however. They entail my own constitution within those places. For instance, my childhood homeworld of the Midwest was a place where, due to harsh winters, we had to struggle to keep grass alive from year to year, had deliberately to plant things, and had to cope with the dryness of the air. Lotion and ChapStick were staples on my bedside table. Upon moving south as an adult, however, I discovered a world of constant humidity and relative warmth. These very physical aspects of the homeworld, then, have cultural impacts that become apparent in the encounter with alienworld. While my current home in the southern United States is familiar because I've lived here for many years, it is still not the measure by which I judge other worlds. The air here is heavy with humidity; plant life is abundant and seems completely random to me. Humid air is abnormal, not what I understand as the norm for weight of air on my skin. There is not the same concern for the cultured lawn as there was in my homeworld in part because it is so much harder to maintain such a cultured lawn without it being overcome by all the other growth. The abnormality of this growth is apparent in my sense that everything is always overgrown here. My own constitution of weeds in this alienworld is very different from my protective response to anything green growing in my homeworld. This world is alien even after all these many years. Normality and abnormality refer to these kinds of differences and merely identify difference and foundation without identifying better or worse. I may have a preference for cold, dry winters, but this does not make them inherently better than warm, humid ones. [3]

In many regards, however, we cannot even come to recognize what is normal in our constitution until we are confronted with what is abnormal. When we encounter alienworld, it is conceived by us as alien in comparison to homeworld. In such an encounter with alienworld, we come to understand some aspect of our homeworld as truly cultural and not as the way the world is in itself. So, we recognize the normality of homeworld when confronted with the abnormality of alienworld. This allows us to begin to recognize what is common about lifeworld. It is lifeworld that makes all else possible for us in terms of constitution of homeworld and alienworld.

This discussion of homeworld and alienworld reveals the history of my experience that is ever present in any new experience I have. That history is never and can never be separable from the world in which I find and have found myself. The homeworld or alienworld is not simply the setting for my experiences; it is fundamental to the way I constitute the world. The objectivity of any world is thus drawn into question. Husserl describes it thus. 'But consider this. Each of us always has his "historicity" with respect to his ego made at home in it. If I am born on a vessel, then I have a piece of my development on the vessel and that, however, would not be characterized as a ship for me in relation to the earth—as long as no unity with the vessel would be produced. It would itself be my "earth," my primitive home' (Husserl 1981, 228).

My homeworld is not isolated, but is wrapped up with a homeworld of my parents and their parents in spite of the differences of generation. These generations have their own homeworlds, but also reach back to a more fundamental horizon and ground of the lifeworld, the core of each homeworld. Again, as Husserl describes, 'All developments, all relative histories have to that extent a single primitive history of which they are episodes. In that connection it is indeed possible that this primitive history would be a togetherness of people living and evolving completely separated, except that they all exist for one another in open, undetermined horizons of earth-space' (1981, 228). The earth-space to which Husserl here refers is the horizon of lifeworld. We turn now to a closer examination of precisely what that lifeworld entails.

LIFEWORLD AS HORIZON AND GROUND

Lifeworld is distinct from these experiences of homeworld and alienworld. The lifeworld is not just empirical, sensuous experience. It is the pregiven, unthematized, natural world upon which both homeworld and alienworld are grounded. 'It is the spatiotemporal world of things as we experience them in our pre- and extrascientific life and as we know them to be experienceable beyond what is [actually] experienced' (Husserl 1970, 138). One may be inclined to rush to the conclusion that this means that homeworld and alienworld are the cultural overlay upon an objective lifeworld. But this is not at all the case. Lifeworld is not an objective world composed of spatial shape, motion or objects subject to gravitational pull. To think such is to already have surpassed lifeworld to a thoroughly scientized world.

The general structure of the lifeworld is bound to its relative being as homeworld or alienworld. In other words it cannot be separated off from any particular cultural world, but is itself not relative. This universal lifeworld a priori is distinct from and provides the ground for an objective a priori that has been established by the idealizing sciences. The sciences are already a 'mediated theoretical accomplishment'. They are dismissive of their own foundations within a lifeworld full of presuppositions in favour of their universal, idealized, geometrized world. Lifeworld is the common structure that all cultural worlds share regardless of their layers of sedimented cultural history. This lifeworld a priori is still a perceptual world whereas an objective lifeworld is not.

Lifeworld a priori is the universal horizon of all experience. Primarily in environmental circles, anyone speaking of a universal, objective world has in mind the kind of geometrized world of the physicist or the environmental scientist. But these approaches are already grounded in the a priori lifeworld that makes such scientific claims about an objective world possible. Such scientists, in order to make their claims or conduct their experiments, rely on the pregiven world of presuppositions that allows for the constitution of this measuring device or that computer model.

Lifeworld is not something we can experience in its wholeness. It is not something we can grasp as an object. There is no position from which we could view the world as a whole. Instead, we can think of lifeworld in terms of its transcendental conditions for the possibility of experience of any kind of homeworld or alienworld or even scientific world. Lifeworld can be conceived as a kind of horizon of experience. Horizon is not something that can ever be made an object of experience, but is nevertheless entailed in any experience, for all experience is horizonal. It is bound by space and time with limits of sense and meaning. This means that the horizon that is lifeworld is pregiven rather than given. It is the very condition of any objects of world as being perceived in their wholeness. Because lifeworld is pregiven, it cannot be described in the same way we might describe a cultural homeworld or alienworld as having particular characteristics or a peculiar sense. Rather, it is the very condition of the possibility of meaning but which itself cannot be made an object. It is horizon and ground of both relative homeworlds and alienworlds.[4]

To think of lifeworld not as object, but as horizon is to recognize it as a way in which something is experienced or revealed. That which is presupposed in the constitution of anything at all is the pregiven lifeworld as horizon of such constitution. It is about a style of constitution of which we are completely unaware and which remains unthematized because it is the very condition of constitution of a cultural world and as horizon of constitution cannot be brought to presence itself. That lifeworld horizon is at the same time a ground of every experience of homeworld or alienworld since it entails the world history of earth that belongs to every living thing of earth.

Any environmental-scientific approach already presupposes this level of ground. Genetic phenomenology provides a method of peeling away layers of cultural sedimentation of homeworld or alienworld to recognize the role of lifeworld as ground. But precisely as ground, lifeworld itself cannot become object over against my investigating subject. It is not a physical body equivalent to what we call planet earth.

UNDERSTANDING LIFEWORLD AS AN APPROACH TO ENVIRONMENT

A phenomenology of nature is important precisely for this particular vantage point that it makes possible. It is a vantage point that allows the theorist to see herself as always involved in the world and responding to the world rather than separating herself from the world and making of the world an object. It also allows us to recognize that in spite of differences of homeworld or alienworld, there are fundamental structures of lifeworld that are pregiven in any worldly givenness. Nature is never object to my subject because we are intertwined in such a fundamental way that my attentiveness to particular elements of nature reveals my embeddedness within a horizon of pregiven nature. So nature and subject are always already interrelated.

The dominance in recent history of viewing our world as a scientific object, as a geometrical sphere floating in space, promotes a style of constitution that views ourselves as somehow other than the world that we evaluate and measure. We fail to see ourselves as able to take a position that questions precisely that homeworld way of constitution. In the natural, scientific attitude, it is simply a truth that the world is an orb in the cosmos that can be measured and analysed, and from which we can ultimately escape. Genetic phenomenology, on the other hand, allows us to see that what we take to be 'natural' about our world is rather a style with a historical origin and many layers of conceptual understanding. For instance, scientific research proceeds on the assumption of objectivity. Particularly with respect to environmental concerns there is often the assumption that our globe is an orb rotating on an axis while revolving around our star, the sun. While there is ample scientific evidence for us to believe this to be true and to treat it as a true claim, it reveals certain prejudices in favour of a kind of second-level theorizing as opposed to more direct experience. As we peel back those layers of scientific presupposition, we begin to see that there are other possibilities for thinking about our world—possibilities that open the way for critique of the dominant scientifico-objective approach that is our homeworld way.

We cannot experience our world as an orb in space. Our experience of our world is conditioned by what we have inherited from prior generations as well as what we ourselves have experienced over the course of the years we have lived. If we are attentive to the way in which scientifico-objective thought has shaped the way we view our world, we can begin to address ourselves differently to that world focusing on how we experience it prior to or outside of that scientifico-objective framework. We can allow our interrelatedness to that world to appear to us in manifold ways. Genetic phenomenology of nature can serve as a partner to the empirical, concrete sciences that are focused on environmental issues of world and nature. In allowing us to peel away sedimented layers of sense, it helps to reveal the presuppositions of our everyday, scientific approach to the natural world and in so doing leaves us prepared for a process of renewal and critique.

Renewal and critique designates a double movement. Renewal takes up traditions of a homeworld such that they do not remain static, but change with each generation. Critique involves questioning those sedimented and inherited convictions. This process, of course, takes place from within the lifeworld traditions that are the horizon and ground of our homeworld. The role of renewal is to make vital again the otherwise concealed sedimented conceptual system which serves as ground of our constitution. This means that we 'reawaken the chain of thinkers, the social interrelation of their thinking, [and] the community of their thought, and transform it into the living present for us' (Husserl 1970, 71–72). On the basis of this renewal, we have a responsibility to critique that chain of thought that we have brought into the present. This means we examine what has been taken for granted, not just by ourselves, but more generally within our homeworld.

With regard to the environment, renewal requires laying bare our inherited traditional ways of thinking about and responding to nature that establish our role as masters of nature, as the ones who can put things right. Critique requires that we draw that approach into question by attempting as far as possible to expose the pregiven elements of our constitution and to move forward with a new kind of thinking. A focus on renewal and critique allows us to think more radically about our relationship to the world as truly and deeply relational and not as either one of dominance or one of inherent value of nature. Instead, we can recognize that because of the layering of sedimentation and the role of the lifeworld, we can take up the sedimented history in ever new and critical ways opening up the possibility of new relations with the world.

Take for example our sense of our world as being traversable in mere hours on an airplane and what that means for how we conceive of our place in the world. It is not difficult to see that this is a peculiar sense of the world that comes from our position within history and within an economically and technologically developed homeworld. This leads to a further sense that we may someday be able to escape our earthly home. Moreover, it contributes to a way of approaching the environment from a perspective that we are able to technologically or scientifically solve any kind of crisis of climate change or loss of biodiversity with which we might be confronted. I do not mean to suggest that all people think this way, but that a general understanding of ourselves that we have inherited is that we have conquered the limitations of space and time through technology and being able to place ourselves 'above' the earth contributes to this kind of thinking. If we engage in genetic phenomenological analysis, we can recognize that this way of thinking about science and technology is peculiar to our homeworld and that before we renew that thinking we should engage in critique of it. Is it a form of thought we want to renew? Is it possible to live radically differently where we could be in harmony with our home as an interrelated element of it rather than as the being that rises above and can control our environment? Do we want to renew the kind of thinking that Heidegger claims arose with Modern technology that the world is here for our use as a resource for our aims?

Part and parcel of this understanding of lifeworld is the recognition that we are through and through creatures of earth and that we are who we are because of our earth-ground and our homeworld. Lifeworld as our horizon serves as our foundation and the foundation for all of humanity. We are fundamentally and thoroughly earthly beings with our constitution of world grounded upon our being earthly beings. Were we from Mars, our constitutional attitude would not be what it is. It would be entirely abnormal by earthly standards. The lifeworld is the world of our normality and anything which differs from that constitutional foundation is going to be viewed as abnormal. This normality and abnormality are not value laden per se, one is neither better nor worse than the other, but one is certainly more foundational and primary than the other. Even if we were to establish a colony on Mars, we would still be earthlings. It is our normality. It is our way of being. We are bound to the lifeworld with all the other beings of this lifeworld and our constitution of any world whatsoever is characterized according to our constitution of this lifeworld, our fundament. In spite of the fact that we live in our own homeworlds we share the lifeworld as horizon and ground of each and every homeworld.[5]

We should not underestimate the power that this shift in thinking has. Genetic phenomenology allows us to see the layers upon layers of tradition and history that compose our way of being, way of doing and way of thinking. With this recognition of our ways as being historical, we come to realize that what we take to be objective truths about our world are grounded in something more fundamental that cannot be objective.

Because lifeworld or nature is pregiven, any theoretical positions or knowledge or understanding that are grounded in that pregivenness cannot be theorized. It can only be given through the homeworld of any culture or constitutional perspective that arises from a particular tradition within a particular temporal horizon. This kind of thinking shatters the scientifico-environmental claim that we can grasp the earth as it truly is in itself. Such a scientific totalizing picture makes way for an attitude of mastery and domination. On the other hand, the genetic phenomenological account provides a foundation for our thinking of ourselves in terms of our ability to see only a part of the picture and to approach our lifeworld in humility and through foundational relationships understanding the ways in which every act of constitution, every act of meaning-making is grounded in a lifeworld a priori that cannot ever be brought to complete transparency. Perhaps we could have a perspective on it someday were we to encounter the Martian alienworld that would allow us to see those things about our earth homeworld that at present are simply too close to us, too intertwined with who we are.

If we consider what it is to see the lifeworld as something that is relational, related to the moon, to the sun, to a cosmos and if we see ourselves as completely composed in relation to that nature which makes us earthlings and not Martians, then we can shed our illusions of an ability to grasp that nature from a godly perspective that conceives of it as a whole and that masters that whole. Genetic phenomenology allows us to recognize that we cannot ever perceive the whole of the world or nature, we cannot have that god's-eye view that allows us to perpetuate an idea of ourselves as masters of the world. We see ourselves instead as humbly dependent upon that lifeworld which is thoroughly intertwined with our cultural world as our unique home.

We also come to see ourselves in our limitations, limited by the horizons of homeworld that are both spatial and temporal and that are grounded upon a lifeworld that can never be made completely transparent. The homeworld-alienworld-lifeworld relationship allows us to see that with such an infinite multiplicity of perspectives and no singular all-encompassing perspective, we are all bound to the lifeworld through our own perspective. This kind of variety takes away any totalizing or dominating point of view presenting us with a picture of ourselves as equally limited by and related to the lifeworld a priori.

The second position that this understanding of lifeworld nature makes possible is an ability to recognize that nonhuman animals and other inhabitants of our world have their own homeworlds and alienworlds that are yet still grounded in a lifeworld for all. The world of the cockroach or the squirrel or the rose bush is alien to me even if I could move in it to some extent, but this does not mean that we share nothing at all. We share lifeworld that makes our homeworlds and alienworlds of experience possible at all. It is not that we humans inhabit a world of culture while nonhuman animals inhabit the natural world. They too have their unique perspective, their own cockroach or squirrel culture, if you will. We all inhabit a homeworld or alienworld supported with the horizons of common lifeworld. Descartes was right in suggesting that our knowledge or rather our experience is relative to the experiencer. The ground and horizon of all experience whether one is cockroach or human is lifeworld or nature.

This nature does not 'exist' independently of homeworld or alienworld. It can only be investigated phenomenologically in order to recognize its deeply interrelational character. We cannot say, then, that world or even lifeworld is anthropocentric. It decentres all insofar as it makes possible the homeworld within which human is centred. But one could also say that in a cockroach phenomenology the cockroach would be centre of its own homeworld. Insofar as both are grounded in lifeworld, both are the same. Perhaps at the level of the cockroach we cannot speak of epigenetics, but certainly at the level of the bonobo monkey or the elephant, we could speak of a sedimented homeworld that is transferred across generations.[6]

We and cockroaches, as embodied beings, are prone to consumption. We eat. To consider what we eat is to think along the lines of homeworld and alienworld. For me, eating bread and cheese is homeworld; for a person raised in China, bread and cheese is more likely alienworld food. For the cockroach, bread and cheese may not be alienworld food, but their consumption of these things is not quite the same as my own. We eat the bread and cheese and it becomes us, becomes our bodies through somewhat different processes perhaps, but still so. The stuff of the world literally becomes my body while the waste from my body becomes stuff of the world. We won't go into the details of how much cockroach waste makes its way into our food, but it does, to be sure just as our waste can be their food. Our embodied natures, vastly different as they are, are part of the a priori lifeworld even as they are moulded and shaped by the homeworld.[7]

Does this foundation make me more inclined to protect or preserve that lifeworld? What it means is that there is a world that is relational, nonobjective, nonsubjective, not merely the result of culture, but not divorced from culture that is the ground of all experience, relational experience, experience that entails the cockroaches, squirrels, and rose bushes with whom my homeworld overlaps.

Lifeworld is not something I can experience except through the filter of homeworld and alienworld. But it is a ground that makes such experience possible and without which there is no common world, no shared place for the cockroach and me. The cockroach is experienced in my homeworld and has a very particular meaning in that world. I am experienced within the cockroach's world as well and although we can't yet claim that the cockroach conceptualizes my existence in its world, it is certainly aware of it as we see when it runs as I raise my shoe to kill it. We have vastly different homeworlds and yet, they are not separate worlds.

We share a place. We share a lifeworld. Is that simply a material, sensual world? No, but neither is it amaterial or asensual. The cockroach is not just a nuisance in my world; I am a threat in its world. We are interconnected, interrelated. Lifeworld makes this possible and that makes a phenomenology of nature necessary.

Particularly with respect to the cockroach or any other living thing with which I share lifeworld, this position allows us to think about our responsibility to and for such beings without needing to establish their level of conscious activity or their ability to suffer. We share a world that is the foundation for and horizon of experience. On that level, the cockroach and I have equal value, equal position. To think through the cockroach within my homeworld is to move to a founded level of constitution where I can recognize that my relation to the cockroach has layers and layers of sedimented history that affect that relationship. I can open myself up to the possibility of thinking that relationship anew since there is nothing 'natural' or 'unnatural' about any of it. The cockroach is not naturally lesser nor am I naturally greater. Historically, perhaps humans have deemed themselves greater for reasons of rationality, but such things have no foundation in a more fundamental lifeworld.

A genetic phenomenology of nature, then, is one that engages us in a process of peeling away the layers of sedimented history of homeworld or alienworld to discover the ground and horizon of lifeworld. This makes possible a twofold approach. It opens us to a position of renewal and critique of our own sedimented presuppositions of meaning, and it allows us to think anew our interrelation to lifeworld and to those beings and things with which we share that lifeworld.

NOTES

1. See for instance, Bryant, Srnicek and Harman 2011.
2. By "our" here I include both human and nonhuman animals.
3. For more on this distinction see Donohoe 2014.
4. For more on this see Anthony Steinbock 1995, particularly chapter 7.
5. One might wonder whether we could adjust to life on Mars and thereby be Martians ourselves. I have argued elsewhere (see Donohoe 2014) that it is not possible for an alienworld to ever fully become homeworld. This does not mean that we could not adapt to some degree, but it would still be measured in terms of homeworld. Across multiple generations this may be something that could wane, but for the generation that traveled from Earth to Mars, that would not be the case and they would still pass along many Earth ways to those born or replicated on Mars such that it would take many generations for Earth ways to no longer be homeworld ways.
6. For support for this see Cynthia Willett 2014.
7. See Simon P. James 2009.

THREE

Natural Phenomena

The Birth and Growth of Experience

Tom Greaves

Phenomenologists have frequently struggled to come to terms with nature. Despite what Ted Toadvine has called the 'inherent paradox of any phenomenology of nature' (2009, 52), namely that, as a philosophy of experience phenomenology seems required to cut nature to the shape and size of human experiential capacity, it is phenomenologists who in recent years have defended nature against critics who find it obsolete, obscurantist and dangerous. Such critics have insisted that even the most nuanced attempts to recover nature for a 'post-nature' world retain and invent new dichotomies that separate us from the very things that we cared about in the first place (Vogel 2015; Morton 2007; Keulartz 1998). Phenomenologists have argued that thinking of nature in terms of *experiential relations* to the world need not involve cutting nature to the pattern of a preformed human understanding. Nature can be thought of not as a set of things that we experience, but in terms of particular features of experience, or even as what inherently resists experience and hence full appropriation by human understanding (Toadvine 1999). In this chapter I will argue that if we are to rescue nature for experience without removing its mysterious and transformative capacity, we need to move through the categories that inform the everyday sense of nature to recover the intensity of the birth and growth of experience itself.

At the same time as admitting that our ideas of 'Nature' have indeed formed a complex set of abstractions that obscure our concrete experience, phenomenologists have also recognized that abstractions of this kind become embedded in our experience. If we want to understand how this happens and perhaps loosen the hold of some of these abstractions, then we need to struggle to recover nature in our own experience. Phenomenologists have developed methods and procedures for recovering vital features of our experience from abstraction and inattention (such as Husserl's 'reduction' of the natural attitude and Heidegger's 'destructuring' of metaphysical thought). In the first section of this chapter I will engage in just such a process. By following a series of contrasts catalogued by one of the first philosophers of experience, David Hume, I will sketch out a rough map of the various abstract conceptions of nature that carve up our experience into preformed categories and eventually drive us beyond experience altogether.

In the remainder of the chapter I try to recover an intense and yet complex sense of nature in experience that refrains from cutting nature down to size. The core idea here is that 'nature' is not something directly or straightforwardly presented to us in experience. Nature is not a phenomenon, nor a set of phenomena. Instead, nature is best thought of as the intensification of features that are to be found in all of our experience to one extent or another.

Phenomenology began with the attempt to understand the *intentionality* of experience, the way that experience relates us to the world. To understand nature phenomenologically, I suggest, we must pay equal attention to the *intensity* of experience. Experience intensifies as phenomena break out in overabundance, while growing into and forming spaces and niches for themselves. We must take account not only of how we relate to the world, but also of the ways that phenomena arise and make room for one another in experience. Intentionality and intensity both have their linguistic roots in the idea of stretching out and straining. If intentionality is our stretching out towards things, intensity is the stretching out of things to accommodate one another as we experience them. The idea takes us back to the sense of nature in the Latin *natura*, with roots in the idea of birth, and in the Greek *phusis*, with roots in the idea of growth. I explore two contrary yet complementary intensities of this kind. On the one hand, there is an intensification of the givenness of phenomena, an abundance or excess of what is given for us to experience that tends to overflow the boundaries of abstract classifications. We might call this *hyper-phenomenal* nature, or the birth of experience. The intense experiences of a sunrise, an opening flower, a nest of wasps overflowing into my living space, or even of the unexpected exposure of the 'inside' of our own bodies when wounded give a first sense of what is meant by such excess. On the other hand, there are open spaces that allow things to appear to us on their own terms, giving things leeway to develop and change unexpectedly. Here we might make use of the idea of 'niche', but not in the sense developed in ecological science. There a niche designates a tightly defined multidimensional set of survival conditions correlating to a particular species. The phenomenological sense of niche points instead towards the spaces and clearings in experience that allow experience itself to grow and change. We might call this *hypo-phenomenal* nature, or the growing space for experience: The trunk of a willow tree split open in a storm allows space for a bright fungal growth, the calm still surface of a lake is broken by surfacing fish with the expanding encircling ripples that follow, a small patch of exposed earth in an urban centre is planted by a guerrilla gardener in the hope of allowing the space to grow to the surprise of unsuspecting commuters. Nature is not directly presented to us in the shape of particular things that we experience, but rather it is the intensification of the birth and growth of experience itself.

DELIMITING 'NATURE' AND NATURE WITHOUT LIMIT

The thought that some things, processes or behaviours are 'natural', while others are unnatural or nonnatural, is deeply embedded into our experience. This idea has long been recognized as historically complex and ambiguous. So, for example, David Hume begins his response to the question of whether we should seek for the ground of moral principles in nature with the salutary remark, 'I wou'd reply, that our answer to this question depends on the definition of the word, Nature, than which there is none more ambiguous and equivocal' (Hume 1978, 473–74). Raymond Williams famously began the entry on 'Nature' in his intellectual lexicon *Keywords* with the remark, 'Nature is perhaps the most complex word in the language' (Williams 1985, 219). An excellent survey of these complexities and the history of various attempts to untangle them can be found in Kate Soper's book *What Is Nature?* (1995). It is important and necessary to trace out these histories and to try to understand how the different senses of the word 'nature' have developed. Nevertheless, beginning with the question What is Nature?, the question that is so obvious and essential to traditional philosophy, may turn out to be part of the problem. The question assumes that nature is some thing or set of things, presented to us in our experience, or perhaps lying somewhere above or beyond our experience. When we begin by asking what nature is, then we assume that we are out to identify some present phenomena and define some specific quality or qualities that makes those particular phenomena 'natural'.

This way of proceeding generally involves setting up contrasts and oppositions between nature and nonnature, the natural and unnatural. Hume identifies three such contrasts: Nature can be opposed to *miracles*; it can be opposed to what is *rare and unusual* (as in when we say 'just act natural'); or nature can be opposed to *artifice*. (Hume 1978, 474) These three contrasts still serve as a good starting point for mapping out the ways in which the concept of nature shapes our experience and eventually takes us beyond our experience. I will examine them in reverse order.

It would be fair to say that it is the last of these contrasts that has attracted the most attention from thinkers concerned with the protection of nature and environmental conservation. After all, aren't many of the things, processes, landscapes and habitats that we are concerned about in one sense or another products of human artifice? And if we extend the thought to include in 'artifice' not only things that have been made or produced by human beings, but things that have been interfered with or influenced by human activity, then we are left to wonder whether there is really anything much at all left on or around the surface of the earth that falls under such a category of 'nature'. This is not simply some rarefied philosophical exercise in categorisation, since whenever we consume something made from 'natural ingredients' or hope to get out and 'enjoy nature', something like this contrast is in operation in our everyday experience. Yet whenever we try to pin down the contrast in precise definitions, we are constantly forced to redraw the supposed demarcation line so that it includes what was unfairly excluded or excludes what we feel was not really the object of our concern.

Simon James makes a valiant attempt to draw such a contrast in a subtle and intuitive way at the beginning of his phenomenological study *The Presence of Nature*:

> A thing is natural to the extent that its current state is relatively unaffected by human agency. On this definition, AstroTurf and human footprints come out as non-natural, since both are produced by humans (the fact that the footprints may not be *intentionally* produced is of no account). Birds' nests and beavers' dams, by contrast, count as natural (they're not the products of *human* agency). Hedgerows come out as natural, for although human beings originally created them, their *current* make-up typically owes more to natural forces (the weather, colonisation by wild plants, the local wildlife) than to human ones. For similar reasons, the Scottish Highlands may be considered natural because, although they were in large part formed through human action (clearances and the like), their current state is, by and large, the product of non-human forces. (2009, 7)

The advantages of trying to draw a contrast between the natural and the nonnatural in these relative terms are clear. We are not forced to make a final pronouncement as to whether each of the objects of our common experience is absolutely natural or nonnatural. Even AstroTurf retains something of its material career before it was affected by human agency. On the other hand, even an Australian coral island of recent origin is likely to be affected in some way indirectly by human agency, especially in the era of acidified oceans and anthropogenic climate change.[1] Without forcing us to stick to an absolute opposition, James's relative demarcation makes sense of a good many intuitive and common judgements about the matter, as he shows in this passage. However, several important worries remain, both with this particular attempt to formulate specific contrast terms, as well as with this way of proceeding in general. We seem now to require some fairly good and reliable background knowledge about the causal processes that have formed each thing in order to experience things as natural and nonnatural. Certainly, experts with historical and scientific knowledge frequently draw into question commonsense judgements concerning the naturalness or nonnaturalness of things. New knowledge can affect our experience and lead us to revise our judgements. But where that knowledge is incomplete, questionable or even entirely absent, our experience itself does not tend to be 'neutral' with regard to naturalness. Even more problematically, we might ask why we should privilege the current state of things in the way that this definition asks us to. I suggest instead that it is not so much the knowledge of past causal constitution, but an extended experience that weaves the past through the present and opens up a space for future growth that makes our experience more or less natural. If we begin not by asking which things are natural, but about the kind of experiences that we are inclined to call natural, then we are no longer trying to carve up the world that we experience, but to encourage and appreciate what is natural in our experience.

That takes us to the second of Hume's contrasts. Does the contrast between what is commonplace and the rare or unusual help us to get a sense of what is natural in our experience? If we are thinking about things in the world that are rare and unusual, then obviously we share a sense of the significance and importance of what is rare and unusual in nature and this sense of significance may be linked to what I describe as the intensity of natural experience. But equally obviously finding or encountering rare and unusual things does not correlate with a common understanding of their naturalness or nonnaturalness. I hear a rare species of warbler on a spring walk, but its rarity does not affect the apparent naturalness of the bird or its song, even if the fact that it is rare may appear to be more or less natural depending on the circumstances. Equally, a rare or unusual event in nature may involve a quite common species or process, but one that is not occurring at the usual time or in the usual way. Furthermore, even if we are thinking of ways our experience itself unfolds, then we may want to describe a common everyday mode of experience as natural. Thus Edmund Husserl described the common attitude towards the world that takes for granted that there are some things that are the case and some things that are not as the 'natural attitude'. The recovery of a sense of nature in our experience that I am advocating does not fit neatly into any of these ideas of the rare and the commonplace. When we attend to our 'natural' assumptions about nature and try to uncover the extent to which they really structure all our experiences, then what we allow to break through in our experience may be at once uncannily familiar and intensely strange.

Perhaps surprisingly, given its apparent reliance on the supernatural, it is the first of Hume's contrasts that shapes the most prevalent modern sense of nature. Here the thought of a contrasting nonnatural category is pushed to the limit. The category of nature as nonmiracle seems to include 'every event, which has ever happen'd in the world' (1978, 474).[2] The contrast is to something utterly beyond experience and beyond the world, and so nature includes everything that there is and everything that has happened. John Stuart Mill, one of Hume's most important successors in the empiricist tradition, extends this idea of nature even further: 'Nature means the sum of all phenomena, together with the causes which produce them; including not only all that happens, but all that is capable of happening; the unused capabilities of causes being as much a part of the idea of Nature as those which take effect' (1904, 8). For Mill this sense of the term, the sum of all actual and possible phenomena, is the primary and most proper sense. For both Hume and Mill this concept cannot tell us anything of interest in evaluative terms, since it includes all virtue and vice, all good and bad, everything proper and improper. Any attempt to persuade us of the goodness of some phenomenon in terms of its naturalness can only be an attempt to smuggle in some ill-conceived prejudice.

I suggested that this has become an increasingly prominent background sense of nature, yet it is by no means the only sense and it is unclear why Mill should insist that it be the primary and proper sense. Both C. S. Lewis and Raymond Williams have offered more subtle readings of the linguistic history of 'nature' that pick up on the contrasts, oppositions and abstractions deployed by Hume and Mill and show that they are often genealogically related in subtle ways. Nature as the sum of all actual and possible phenomena is what Lewis calls 'the whole show', 'Nature with a capital, *nature* (d.s.), or nature in the dangerous sense' (Lewis 2013). It is a dangerous sense because it is the sense that is most likely to intrude where it is not required. It has become the assumed background or primary sense of a term that has a much richer and more complex historical and experiential sense. Interestingly both Lewis and Williams concur that nature in something like this sense has its roots in the philosophical project of asking for the essence or nature of things. Linguistically that sense of the 'nature' of a species or kind of thing, its essence, or that which makes it what it is, arises earlier and was for a long time far more common than the senses of a particular set of natural things contrasted to nonnatural things, or the sense of the sum of all phenomena. The nature of a thing may seem to be far removed from Nature as the sum of all things. Perhaps it seems far more real and concrete. I experience all kinds of things, but Nature as the sum of all things is forever beyond my experience. Nevertheless, to ask after the nature of a thing, to ask for its 'essence' or whatness, is already to set foot on the path of abstraction that leads us to ask after the nature of 'the whole show' and to ask and answer the question 'What is Nature?' (Lewis 1960, 36; Williams 1986, 220). To set out on the phenomenological path requires us to bracket that question and to describe instead *how* our experience can and does become intensely natural when we focus not on what is presented to us, but on the birth and growth of experience itself.

The Birth and Growth of Experience

Attempts to carve up phenomena into the natural and the nonnatural, together with the limiting thought of nature as the set of all actual and possible phenomena, certainly affect our experience. These structuring thoughts are generally unavoidable and frequently necessary. However, in the remainder of this chapter I will make the case that there remains a complex sense of nature in experience that these structuring thoughts do not and cannot capture. Nature as the intensification of experience, I contend, requires two contrary yet complementary moments in which phenomena break forth and are given space to come into their own, the birth and growth of experience.

In the work of phenomenologists there are many hints and descriptions that help to draw out this phenomenal sense of nature. Martin Heidegger above all insisted that it was the contrary yet complementary tendencies in our intense experience of phenomena that lets beings come forth and come into their own. His configuration of those tendencies shifted and developed in the course of his work, from the well-known 'strife of world and earth' in his essay on *The Origin of the Work of Art* to the folding of 'earth' and 'sky' into the fourfold of earth, sky, mortals and divinities (2002; 1993, 343–64).[3] In one late essay Heidegger suggests that we should try to think through the basic tendencies of appearance in terms of 'presencing' and 'clearing' (1993, 427–49). The thought can be concretized as a sense of phenomenal nature that draws out the birth and growth of experience. Crucially, it is the vital tension between presencing and clearing, birth and growth, that naturalizes phenomena.

Heidegger tried many times to describe the sense that *phusis* had for the ancient Greeks, the term that would later be translated as 'nature'. In a seminar in 1969 he described it as 'the *overabundance*, the *excess* of what presences.' (2003, 38; cited in Capobianco 2014, 38) This sense of excess does not take us above and beyond our common experience, but is an intensification of that experience. A good example can be found in the following passage from Aldo Leopold:

> Everybody knows, for example, that the autumn landscape in the northern woods is the land, plus a red maple, plus a ruffed grouse. In terms of conventional physics, the grouse represents only a millionth of the mass or the energy of an acre. Yet subtract the grouse and the

> whole thing is dead. An enormous amount of some kind of motive power has been lost. (1949, 122)

The intensity of this experience comes not from the isolated focus upon a red maple or a ruffed grouse, nor from the vague encompassing sense of the landscape shrouded in mist, although each of those are moments in the overall experience. The intensity comes from the tree and the grouse experienced as integral to the landscape and yet somehow also in excess of it, pointing us to the excess and overabundance of what has already become present and what may yet become present. The intensity of birds breaking into flight, hidden cries or breaking branches, the smell of encroaching autumn, are all folded into overabundant nature. This is not a question of our finding a single focal point around which the experience organizes itself. It is the excessive sense of addition, of the maple and the grouse, that intensifies the experience—excessive because they are of course already part of the land and yet present themselves anew here. Dissatisfied with the inability of modern physics or scientific ecology to describe such an experience, Leopold goes on to invoke a philosopher (apparently Kant) and calls the grouse the 'numenon' [*sic*] of the forest. Yet what he has described is not at all the thing simply thought in itself and forever beyond experience, which is what Kant understood by noumenon. It is instead the excessive and overabundant birth of experience from out of itself.

It is not only in classic 'nature scenes' such as that evoked by Leopold that we experience the birth of experience. This is a sense of nature that does not oppose nature to the human. Perhaps most obviously in giving birth there is an experience of the intense overabundance of nature. Such natural intensity can be retained and perhaps even rescued by various kinds of assistance that in common parlance do not make for a 'natural' birth. There are many other intensities of excess and abundance that we experience every day. Think of the sharp sting and bright flowing blood of a wound, whether inflicted by a thorn or a knife. Or the breaking into speech that comes with almost every human encounter.

The French philosopher Jean-Luc Marion has a keen sense for this feature of phenomenality. He designates it with the term 'saturated phenomena' (Marion 2002a, 2002b). If we begin with and remain with the enigma of the pure *givenness* of phenomena, then the excess of what is given comes to the fore, overwhelming our attempts to structure and confine that excess. This is very close to what I am describing as the birth of experience from out of itself. Bruce Foltz and Christina Gschwandtner have both argued that Marion's notion might be fruitfully employed to think about natural phenomena, as we move away from a notion of nature as a self-subsistent and fully intelligible system and try to regain a sense of nature's 'other side', the givenness of phenomena (Foltz 2004; Gschwandtner 2014b). One important point emphasized by Foltz is that what is at stake here is not simply the identification of 'certain extraordinary phenomena' (333). It is not the red maple and the ruffed grouse in splendid isolation that exclusively display natural givenness. On the contrary, saturated phenomena are icons, paradigmatic of phenomena as such, precisely because they display to a hyper-degree the givenness that is the very definition of all phenomena.

Nevertheless, if the saturated phenomenon *gives itself* and that givenness is not countered by a leeway or niche that is never completely filled, it tends towards a self-sufficiency and completeness in its own right. Gschwandtner describes this concern as the need to recognize 'degrees of givenness' (2014a), although I would suggest that it is not the givenness as such that comes in degrees, but the specific intensity of givenness that finds its own leeway in singular experiential contexts and situations. The desire for complete saturation and inattention to natural phenomena may be linked to Marion's attempt to move from a phenomenology of givenness to a theology. The excess of givenness becomes an apotheosis beyond epiphany. Saturation becomes an unnatural fullness and completeness of what gives itself that leaves little or no leeway for growth. Think, for example, of what we often understand by 'natural' light. It is not a light that spotlights what is right in front of us, holding our attention fixed just there. Neither is it a light that saturates the scene, like fluorescent tube lighting in a bare whitewashed basement. A natural light reveals and plays around what it reveals, holding out the promise of quite new revelations and the withdrawal of those that first came into view. It may be the light of the sun and moon, a lightning strike and resultant bush fire, or fireplace around which we gather.

Understanding the naturalness of phenomena in terms of overabundant givenness helps to undermine the idea that what is at stake in the experience of nature is an attempt to bridge a gap between facts and values. Experiences of naturalness are not divided up in that way. At the same time we must be wary of attempts to draw straightforward normative conclusions from the recognition of naturalness in our experience. Fluorescent tube lighting tends to saturate a scene and reduce the space for phenomenal growth in that scene, but its use may help to reduce electricity demand, and in the 'big picture' potentially allow phenomena to show themselves and give them leeway to grow, without being immediately plugged into the grid. Foltz suggests that an assault on nature as givenness is evident in numerous modern practices, from factory farming and genetic manipulation to abortion (2004, 339). That suggestion not only ignores the very significant contextual differences between these experiences and practices; it also obscures the complex ways in which they are historically and politically intertwined. The recovered experience of givenness must not be made into an excuse for unreflective immediacy in our ethical and political judgement. The need for leeway in which the living experience of humans and other living beings can grow and unfold gives another dimension to the experience of natural phenomenality than one focused on pure givenness. Even then, no phenomenology of nature will solve all our ethical and political dilemmas in one go, even if it can return us to a sense of the interconnectedness of these problems and why they seemed so important in the first place.

Turning now to the 'growth' of experience, the crucial point to emphasize here is that growth is not opposed to withering, decay or death. We are familiar with spaces that are given for things to grow from the death, decay and waste, such as the fungal growth in a split tree trunk or a growth of maggots in a carcass, or an orchid springing up in the sandy ground of a building site. Such spaces are formalized by ecological science in the concept of niche, formed by the complete set of survival conditions for a species. But how are we to make sense of this thought when we move from the set of phenomena that we find before us to the ways in which phenomena appear to us in and through experience? Where do we find room for niches in experience?

Such niches open up and close over experience at every turn. They are hardly ever noticed, because they are the making of a space for something to appear. They are familiar from powerful and overwhelming experiences, like the lull before a storm breaks. They can also be found in less dramatic form in the intimate crevices and nooks that harbour and enclose experiential potential, or the paths and tunnels that connect elements of a scene in ways that are not themselves revealed in the experience. Niches of experience allow phenomena to come forth, grow, gather and disperse, to show themselves in their intense excess and overabundance. At the same time what appears in them must leave some leeway so that they are not filled entirely with what is present and allow no room for growth. The niche allows phenomena to grow and that growth in turn maintains a niche for itself.

Let us return to Leopold's landscape in the northern woods. Once autumn turns to winter, the red maple withdraws from prominence. The brilliance of its autumn foliage is made all the more startling by this withdrawal. The ruffed grouse has burrowed in the snow. Yet the landscape is not 'dead', even if our initial prejudices might suggest as much. It is not dead in biological and ecological terms, but neither is it dead in experiential terms. The grouse might burst from its snow hole at any moment, leaving traces of its sudden excess of flight as wing beats marked in the powder. If it remains hidden it leaves room for anticipation of the coming season, the drumming and displays of its courtship that will come with the abundance of spring.

The growth of experience does not simply involve increase and expansion. As with any growth, if pure increase outstrips the creation of a niche for growth, then that increase will be self-limiting and often self-undermining. Growth here, as elsewhere, involves growing self-emergence that allows leeway for further unfolding, and that in turn entails withdrawal, decay and demise. The growth of experience I am pointing to here can be helpfully contrasted to Hannah Arendt's notion of 'natality.' Arendt fixes natality as one pole in an opposition that she claims bifurcates the 'general condition of human existence: birth and death; natality and mortality' (1998, 8). Furthermore, she focuses our attention on the *newness* that is clearly a significant feature of natality in a way that tends to mark it out as its sole and essential feature. What is lost here is a sense of the growth that has already brought us to the point of each birth of 'newness', birth that is never an absolute break from what came before, but its intense culmination. Lost too is the withdrawal, decline, decay and mortality that make room for newness. Natality and mortality are not opposing poles of the general human condition, but rather the contrary yet complementary tendencies that bring forth what is new in the spaces opened up by withdrawal, decay and death. Niches in experience are brought about by the passing away of abundance. They are also brought about and shaped by the growth and unfolding of what appears in them. This is the natural appearance of phenomena. It is not a matter of particular qualities that things display once they have appeared, but of how they bring about an intensification of experience as they appear. All experience retains something of these natural capacities for birth and growth, but there are some practices and ways of life that acknowledge and foster them, while others are oblivious or obfuscatory, which is why natural experience inevitably retains a historical and political dimension.

Experience is never a blank sheet waiting to be filled with whatever impressions come to make their mark, as empiricist philosophers like Hume thought of it. If we think of it like that then 'nature' can only ever be a particular set of impressions, or perhaps the 'blank sheet' itself, forever beyond experience. Such an image of experience is reminiscent not of the niches worked into a landscape by the birth and growth of the creatures living there, but of a clear-cut forest that momentarily seems to lose all capacity for birth and growth. Everything is exposed and so there is nothing left to reveal itself. In such clear-cut space what does appear will tend to take up all of the available space, and in experience monopolize all of our attention. The northern landscape without the red maple and the ruffed grouse is dead, but a clear-cut forest will provide no refuge for grouse and will quickly be overrun with maple.

What should be clear by now is that a phenomenology that struggles against conceptual contrasts to recover a sense of the birth and growth of phenomena does not allow us to derive straightforward normative consequences from its understanding of the naturalness of phenomena. It has that in common with the thought of empiricists like Hume and Mill. But unlike those empiricists, it insists that 'nature' is not simply a more or less adequate conceptual tool for carving up the world or for designating that world as a whole. There is also an intense naturalness in the vast multiplicity of ways in which that world is born and grows in and through experience. Such experience is never cut off from the complex political, economic and ecological histories that feed into it. Leopold's northern woods are not immemorial, even if their excessive abundance grew within them for a long time before it was born into his experience. Any such experience of natural phenomena opens directly onto an aesthetics and ethics that does not elide such histories. The intensity of experience insists that nature is not simply what has always presented itself or what is currently present, but makes itself felt in the birth and growth of phenomena.

NOTES

1. The example comes from Marx and Engels (1974, 63) who already recognized that there is probably no longer any absolutely 'pure' nature to be found anywhere on earth.
2. '[E]*xcepting*', Hume continues with apparent irony, '*those miracles, on which our religion is founded.*'
3. Of the many attempts to think through the condensed and elliptical formulations of the 'fourfold' in Heidegger's late essays, Andrew Mitchell's (2013) reading is one of the clearest and most illuminating. His reading of 'earth' as the sheer intense presencing of phenomena and 'sky' as the clearing that allows for that presencing, has informed and inspired my attempt here to elaborate on the contrary yet complementary tendencies of phenomenal nature.

FOUR

Phenomenology and the Charge of Anthropocentrism

Simon P. James

In discussions of environmental thought, 'anthropocentric' tends to be used as a term of abuse—not as abusive as 'sexist' or 'racist' perhaps, but fighting talk nonetheless. Yet what exactly does 'anthropocentrism' mean in such contexts?

The term is often used to denote a general thesis about what has value. To be anthropocentric, on this conception, is to suppose that the nonhuman—or, if you like, more than human—world has value only because, and to the extent that, it serves human interests (see, e.g., McShane 2007b, 170). So, for instance, when Aristotle (1995, 23) implies that all nonhuman lifeforms were 'made by nature for the sake of men', he seems to be espousing this sort of anthropocentrism. In some contexts, however, 'anthropocentrism' is used to denote a general thesis about what sorts of things exist. It is used, that is, to pick out a family of ontological, rather than axiological, positions. To put the point roughly and figuratively, a position is taken to be ontologically anthropocentric if it fails to allow for the existence of anything beyond the sphere of the human. For example, when environmental thinkers such as George Sessions (1995), Holmes Rolston III (1997) and Eileen Crist (2004) complain that what they call postmodernism is anthropocentric, they are using the term in this ontological sense. It is anthropocentric, they claim, for postmodernists to hold that what we usually take to be the nonhuman world is in fact nothing more than a cultural artefact, a social construct, the outcome of negotiations between interested parties, or in some other way a reflection of what goes on in the human world.

Ontological anthropocentrism is thought to have several shortcomings. First, many forms of it are believed to be false (and some forms of it certainly are false). Second, merely endorsing some kinds of ontological anthropocentrism seems, arguably, to indicate a flaw in one's character. (Witness Bertrand Russell's claim that 'greatness of soul is not fostered by those philosophies which assimilate the universe to Man' [1967, 92] or Thomas Nagel's statement that idealists betray 'a lack of humility' [1986, 109].) Third, ontological anthropocentrism might be thought to encourage the morally dubious position of axiological anthropocentrism. For instance, it might be supposed that although the nonhuman world is valuable in its own right, one cannot value it as such if one regards it as nothing more than a reflection of what goes on in the human world (see, for example, Coates 1998, 185).

Speculative realists such as Quentin Meillassoux argue that phenomenology is inherently anthropocentric in an ontological sense.[1] Phenomenologists, they maintain, focus all of their attention on how reality discloses itself to us human beings, dismissing speculations about 'the autonomous reality of the nonhuman world' (Sparrow 2014, 19; cf. Carman 2008a). Furthermore, some speculative realists claim that this ontological anthropocentrism should worry those who are concerned about environmental problems. For example, Levi Bryant, Nick Srnicek and Graham Harman associate phenomenology with 'an anthropocentric stance towards nature'. It is, they maintain, unclear whether any such approach will help us to face up to 'the looming ecological catastrophe' (Bryant, Srnicek and Harman 2011, 3–4). Tom Sparrow, another speculative realist, agrees. He claims that 'the antirealist reduction of reality to human reality' which one finds in the works of phenomenologists goes hand in hand with 'the anthropocentrism that . . . fuels the current climate crisis' (2014, 19).

In the following, I draw on the works of Maurice Merleau-Ponty to challenge these accusations. Phenomenology, I contend, need not be ontologically anthropocentric.

THE ASEITY OF THINGS

We are human, so whichever way we turn we see things in distinctly human ways. For example, when I visit Yellowstone National Park, I see it through the eyes of a white, middle-class, educated British man. My perceptions of the place have been influenced by a range of cultural factors: by the writings of Henry David Thoreau and John Muir, by the paintings of the Hudson River School, by countless wildlife documentaries I have seen, by various notions about wildness, colonialism, masculinity, private property and the American West. And at a deeper level, my impressions of the park have been shaped by the fact that I think and feel like a human being rather than, say, a wolf, and that I have a human body rather than four padded paws, a tail and a supersensitive nose. This holds true, moreover, whether I am actually perceiving the park or simply imagining it. It holds true even if I am merely thinking about Yellowstone without calling any sort of mental image to mind. The general upshot is clear. Although I might like to think that in visiting Yellowstone I am confronted by a starkly nonhuman world in which nonhuman things go about their nonhuman business in their own nonhuman ways, my impressions of the place have in fact been thoroughly shaped by various 'human' factors.

The general thought here is that:

A. What we take to be the world, its nonhuman parts included, is necessarily and thoroughly shaped by various 'human' factors.

Suppose, for argument's sake, that (A) is true. One response would be to suggest that:

B. The world, its nonhuman parts included, is nothing more than a combination of various 'human' factors.

(B) is clearly an ontologically anthropocentric conception in the rough sense sketched above. It clearly does not allow for the existence of anything beyond the human. According to it, the world may be likened to an onion. Onions don't have cores: one can strip away layer after layer until one ends up with nothing at all. Similarly, if (B) is true, then stripping away all the 'human' factors that shape our impressions of the world would leave us with nothing at all.

So far as I am aware, no phenomenologists would endorse (B). Merleau-Ponty, for his part, certainly would not. On the contrary, he insists that the world, its nonhuman parts included, is *not* merely what we humans make of it. Thus he chastises transcendental idealists for regarding the world as 'immanent in consciousness', thus occluding the 'aseity of things' (1962, xv-xvi). And he expressly rejects the claim that 'the world is constituted by consciousness' (432). The truth of the matter is, he insists, that the world 'is always "already there" before reflection begins—as an inalienable presence', one which consciousness can neither 'embrace nor possess' (vii, xvii). Indeed, this, he suggests, can be borne out in perception. If we 'suspend our ordinary preoccupations' and regard any particular thing with 'disinterested attention', we find ourselves confronted by a 'hostile and alien' presence, one which harbours a 'non-human element' (322).

It might be objected that the mere fact that Merleau-Ponty made these realist-sounding claims fails to prove that his phenomenology does not entail the truth of (B), for he might have made them *despite* his phenomenological commitments.[2] But that is not the case. Even writers sympathetic to speculative realism concede as much.[3] In sum, Merleau-Ponty is not rationally committed to the particular variety of ontological anthropocentrism implied by (B). He need not—and does not—suppose that the world is nothing more than a projection of the human mind, a social construction or some other combination of 'human' factors.

PHENOMENOLOGY AND CORRELATIONISM

Merleau-Ponty would not endorse (B); his phenomenology is not ontologically anthropocentric in that sense. But is it ontologically anthropocentric in some other sense?

We have seen that to endorse (B) is to buy into a certain form of ontological anthropocentrism. Furthermore, one can certainly reject (B) without endorsing any sort of ontological anthropocentrism. It is, however, a further question whether one can consistently reject (B) and nonetheless be committed to some form of ontological anthropocentrism.

Meillassoux would say that this is possible. In his view, two sorts of ontological anthropocentrist would reject (B). *Weak correlationists* follow Kant in supposing that although there *is* a way that the world is, independent of any 'human' factors, finite beings like ourselves could never know what this world 'in itself' is like. *Strong correlationists*, by contrast, would say that although one might be able to conceive of worlds for nonhuman subjects, the notion of a world that is not a world for any possible subject makes no sense. For the strong correlationist, it is therefore pointless to speculate about how the world might be in itself.

Endorsing strong correlationism does not commit one to the truth of (B). One can be a strong correlationist and yet deny that the world is simply what we make of it. The strong correlationist can accept that we find ourselves inhabiting a world which resists our attempts to comprehend it with our minds or shape it with our hands. But she will reject speculations about how this world might be in itself. For her, any world is always 'lit up' in terms of the concerns, moods, etc. of those beings for whom it is (or would be) a world.

The Merleau-Ponty of *Phenomenology of Perception* seems, on the face of it, to be a strong correlationist. True, he signals his rejection of (B) by claiming that the world is 'always "already there" before reflection begins', yet the fact that he places 'already there' in scare quotes seems to indicate his continued allegiance to correlationism (1962, vii; cf. Sparrow 2014, 45). To be sure, he says that when we suspend our ordinary preoccupations, we find ourselves confronted by a hostile and alien thing. But he immediately adds that it 'can never be actually *in-itself* because its articulations are those of our very existence, and because it stands at the other end of a gaze or at the terminus of a sensory exploration which invests it with humanity' (1962, 320). It is merely an '*in-itself-for-us*' (322). Once again, underneath the realist-sounding language, Merleau-Ponty seems to remain correlationist. For him, the world is, it would seem, a 'human' world (cf. Cooper, this volume).

For Merleau-Ponty, then, what we perceive is indelibly coloured by our distinctly human concerns, 'overlaid', as he puts it, 'with anthropological predicates' (1962, 320). And he suggests that this holds true, also, of anything we might be able to *conceive*: '[W]e cannot conceive anything', he claims, 'which is not perceived or perceptible' (320). This is why he dismisses speculations about what things are like in themselves, divorced from any relation to a subject. It is why he famously took issue with A. J. Ayer's claim that the sun existed before human beings (see Toadvine 2014). And it is why he takes the line he does in the following passage from *Phenomenology of Perception*:

> To our assertion . . . that there is no world without an Existence [i.e., a being for whom there is a world] that sustains its structure, it might have been retorted that the world nevertheless precedes man. [But] what precisely is meant by saying that the world existed beyond any human consciousness? An example of what is meant is that the world originally issued from a primitive nebula from which the combination of conditions necessary to life was absent. But every one of these words, like every equation in physics, presupposes *our* pre-scientific experience of the world, and this reference to the world in which we *live* goes to make up the proposition's valid meaning. Nothing will ever bring home to my comprehension what a nebula that no one sees could possibly be. (1962, 432)

Merleau-Ponty's phenomenology, as presented in *Phenomenology of Perception*, seems therefore to be an example of strong correlationism.[4] Speculative realists such as Meillassoux would say that, as such, it exemplifies the anthropocentrism which marks so much post-Kantian thought. Be that as it may, even if the position developed in *Phenomenology of Perception* really is ontologically anthropocentric, it is a further question whether the same may be said of Merleau-Ponty's other contributions to phenomenology, still less those of Husserl, Heidegger, Marcel, Sartre, et al.

The first thing to note, in addressing these further questions, is that it is misleading to refer, as I have just done, to Merleau-Ponty's *position*. Neither Merleau-Ponty's phenomenology, nor any other form of phenomenology, is best conceived of as such. Phenomenology is, rather, an *approach*. The key question, then, is whether adopting such approaches must lead one to ontological anthropocentrism. And, as David Morris (2013, 327) argues, before concluding that the answer must be 'yes', one ought to look to see where the paths of phenomenology do in fact lead. In the remainder of this chapter, I shall argue that they can lead one beyond the sphere of the human.

TOWARDS THE FLESH

Return to Merleau-Ponty's account of our perception of things. The thing's articulations, recall, 'are those of our very existence' (1962, 320). So, for instance, when I turn my attention to the semi-precious stone—a tiger's eye—sitting on my desk, what I perceive is conditioned by my memories, my cultural context and my body as well as by a host of other 'human' factors. Since it's a hot day, the stone reveals itself to me, even before I touch it, as having a pleasantly cool surface. It reminds me of visiting a seaside town with my mother many years ago. Furthermore, my perception of the stone—both what I perceive and how I perceive it—is shaped by the norms of the various cultures to which I belong. The stone stands out in my field of perception as mattering, yet as neither expensive nor sacred. And as Merleau-Ponty would emphasise, my perception of the stone is also conditioned by my body. Indeed, this remains the case even when I am not touching the thing. I don't need to reach out to know that it is within reach: I *see* it as being within reach. I do not need to make contact with my fingers to feel its hardness or smoothness: I *see* it as hard and smooth.

Much more could be added here; however, no observations of this sort would allay the worries of Meillassoux et al. For, to the speculative realist, any phenomenological account of the perception of a stone—or of anything else—will continue to portray what we take to be part of external reality as a correlate of the body-subject. To be sure, the phenomenologist uses what Sparrow (2014) calls a 'rhetoric of realism'. She tries, through a clever use of words, to evoke the presence of an external world. Yet the speculative realist will insist that the basic position remains correlationist. The stone remains a stone-for-us.

Even so, the phenomenologist's description *points beyond* ontological anthropocentrism. Return to the tiger's eye. I reach out to touch it—but what exactly is going on here? My hand is not an instrument I use to explore the thing, for talk of instruments implies that the agency in perception lies entirely with me and a more thorough phenomenological investigation reveals that that isn't the case at all. Rather, the stone's surface invites my touch. It 'sets a kind of muddled problem for my body to solve', writes Merleau-Ponty; it 'summons' me and my body 'welcomes' it (1962, 214, 318; 1964, 164). I find that my hand has already adopted the posture required to perceive the stone's surface. The muscles in my fingers are relaxed; they extend towards the stone, anticipating smoothness.[5] They have already come alive as sentient, and the stone's surface as smooth, even before they make contact. Just as my fingers in this sense create the stone's smoothness, so that smoothness gives birth to my fingers as sentient organs. So it is not as if I use my fingers to explore the stone. No, in a sense, I am touched by the stone, just as it is touched by me. As Merleau-Ponty puts it, in such cases 'it cannot be held that one acts while the other suffers the action, or that one confers significance on the other. Apart from the probing of my eye or my hand, and before my body synchronizes with it, the sensible [i.e., in this case, the stone] is nothing but a vague beckoning' (1962, 214).

It is these ideas which point the way beyond ontological anthropocentrism. Granted, the *content* of perception, *what* I perceive, is indelibly coloured by my human concerns. Like everything else I perceive, the tiger's eye continues to display 'the human face it acquires in a human gaze' (Merleau-Ponty 2002, 54). Yet the perception itself is not entirely my doing, for the agency—if that is right word—lies partly with the thing (cf. Toadvine 2009, 86). As Merleau-Ponty puts it, the thing *invites, summons, welcomes* or *beckons* my hand.[6]

In *Phenomenology of Perception*, Merleau-Ponty maintains that perception presupposes a basic accord between the pre-personal self which, as he puts it, 'perceives in me' and the surrounding things by which this 'natural self' finds itself invited, summoned, welcomed or beckoned (1962, 215, 440). In his later works, he adds that that accord is possible not only because the self is situated *in* the world, but also because it and the world it inhabits, perceiver and perceived, are of the same 'stuff'—not matter or mind, but a common 'element' which he calls 'flesh' (*chair*) (2003, 218; 1968, 139, 146). That suggestion might conjure up a mental image of a relation between two distinct relata, a fleshy perceiver and a fleshy world it inhabits. Yet in his final, unfinished work, *The Visible and the Invisible*, Merleau-Ponty makes a radical suggestion: that what might, in other circumstances, be called an act of perception should be conceived of as something like an event from which both perceiver and perceived are abstractions: a 'coiling over' or 'intertwining' of a single flesh (1968, ch. 4).

BEYOND THE HUMAN

There is not enough space, here, to give a full account, still less a defence, of Merleau-Ponty's notoriously abstruse account of flesh. Suffice to say that there has been a great deal of debate about how that account should be interpreted. Some claim that it marks the man's rejection of phenomenology for metaphysics—a move which not all applaud.[7] Others disagree (e.g., Madison 1990, 32). Some maintain that the account of flesh signals Merleau-Ponty's rejection of correlationism; others hold that it remains thoroughly correlationist.[8] And of course quite a few people think that Merleau-Ponty's claims about flesh are mostly false. Still, whatever the answers to these questions, two things are clear. First, even if (which is by no means clear) the account of flesh marks a decisive break with the phenomenological tradition, it is clear that it was arrived at *by means* of a phenomenological approach. Second, even if that account remains correlationist, even if, in fact, it cannot ultimately be justified, it is clearly not ontologically anthropocentric.[9] In *Phenomenology of Perception*, recall, things are described as being 'overlaid with anthropological predicates' (1962, 320). In *The Visible and the Invisible*, by contrast, Merleau-Ponty insists that he does 'not mean to do anthropology, to describe a world covered over with all our own projections, leaving aside what it can be under the human mask' (1968, 136). Instead the reader is invited to think of flesh as 'the concrete emblem of a general manner of being': a way of being which, to be sure, expresses itself in what I call my perception, but which extends beyond this to encompass 'the whole of the sensible of which [the body] is part, and . . . the world' (1968, 147, 138). As such, it is thought to exist 'in other fields' (144)—not just in the stone's welcoming of my hand, but in the foliage inviting the elephant's trunk or the river the kingfisher's dive. In such cases, too, sensible calls to sentient—these are also events of nature, intertwinings of flesh.

To repeat: With these thoughts, the focus has shifted away from the *content* of perception—which presumably continues to display the human face it acquires in a human gaze—to what might be described (though imperfectly) as the *act* of perception.[10] That is to say, the guiding question is no longer 'What is perceived?' but 'Who (or what) is perceiving?' And in addressing *that* question, Merleau-Ponty found himself drawn towards a conclusion prefigured in the works of thinkers such as Schelling, Schopenhauer and Coleridge: that what we misleadingly call *our* experience is the manifestation of some wider and deeper upsurge of being (see further, James 2009, ch. 5).

In sum, then, it was precisely phenomenological inquiry which eventually led Merleau-Ponty away from ontological anthropocentrism. The claim that all forms of phenomenology are ontologically anthropocentric ought therefore to be rejected. Perhaps phenomenology has some essential flaws, but ontological anthropocentrism isn't one of them.

NOTES

1. Meillassoux himself espouses what he calls 'speculative materialism' (Harman 2011, 168). For present purposes, however, 'speculative realism' and 'speculative materialism' may be regarded as synonyms.

2. Cf. Tom Sparrow 2014, 50–51. Incidentally, Sparrow isn't being entirely fair here. After all, isn't what eminent self-proclaimed phenomenologists *say* the best guide to what phenomenology *is*? It would surely be unfair to insist that phenomenology is essentially anti-realist and then to dismiss any realist-sounding statements phenomenologists make as contrary to the spirit of the tradition.

3. In fact, even Sparrow admits this. 'Merleau-Ponty's lesson at the end of the day is', he writes, 'that the world we perceive is always more than what we perceive, and it is perception that discloses this to us' (2014: 44).

4. I say *seems* since that claim is open to question. Consider Merleau-Ponty's account of the thing-in-itself-for-us. It is not supposed to convey scepticism—the thought that although there seems to be more to the thing than we can perceive there

might in fact be nothing more to it. Nor is it as if we encounter the thing and *then*, by means of inference or some other mental operation, conclude that it exists independently of us. No, its independence from us is partly constitutive of the thing as perceived. This may be one of the points Merleau-Ponty means to convey by speaking of *horizons*. Just as a distant horizon both lies within and points beyond my visual field, so the thing both lies within and points beyond the correlation. For an illuminating discussion of this difficult issue, see Toadvine 2009, ch. 2.

5. Granted, there is room for error here. Perhaps the smooth-looking surface turns out to be rough. But error is only possible because there is a basic harmony between perceiver and perceived.

6. Don't let these words mislead. This basic accord between sentient and sensible is also the condition for the possibility of one's being repelled by the things one encounters, as when one's hand spontaneously recoils from something sharp or slimy.

7. For instance, Taylor Carman (2008b, 241n38) expresses doubts about the trajectory of Merleau-Ponty's later, increasingly ontological inquiries. By contrast, Bryan Bannon suggests that in his later works Merleau-Ponty was moving in a more promising direction: away from 'the traditional project of phenomenology' and towards a Whitehead-inspired 'speculative process philosophy' (2014, 185n40). I take no stand on this issue here.

8. The former view is defended in Madison 1973, 205–7 (cf. Merleau-Ponty 1968, 131). By contrast, Sparrow (2014, 49, 81, 155) argues that the account of flesh remains correlationist. I suspect that Meillassoux himself would claim that in his later works Merleau-Ponty 'absolutizes' the correlation (see further, Meillassoux 2008, 37).

9. And even if it were, it would be a further question whether, as Sparrow et al. seem to assume, ontological anthropocentrism is necessarily at odds with environmental concern. A comparison with artworks may be useful here. The world of art is clearly anthropocentric in the sense that claims about art make no sense when they are entirely abstracted from references to the attitudes, concerns, etc. of beings, like us, which are capable of appreciating art. Yet no one would claim that since, say, Michelangelo's *David* is anthropocentric in this sense, the statue may legitimately be used as a makeshift drying rack for wet laundry or broken up for ballast. Likewise, it is not clear why someone who believes that the natural world is a 'human' world must be inclined to regard it as a collection of mere stuff to be used in any way she sees fit. (I would like to thank David E. Cooper for suggesting this point to me.)

10. Merleau-Ponty expresses his misgivings about talk of perceptual acts at 1962, x–xi and 1968, 244.

FIVE

Nature, Meaning and Value

Bryan E. Bannon

For those of us concerned with the preservation of the natural environment, there are few claims more intuitively obvious than that nature itself possesses value and that the existence of wild nature enriches our lives in meaningful ways. The feeling of connection between self and environment expressed in these claims is used as a support for the further argument that this value ought to be preserved, and something is amiss when we ignore this value in favour of a narrow set of human interests. Regardless of whether one criticizes the selfishness of such behaviour or the violation of nature's integrity, the validity of the evaluative claim rests upon the acceptance of nature's value.

For this reason, a great deal has been written trying to establish that nature possesses an intrinsic value independent of any value we ourselves derive from our use or enjoyment of the natural world. There have been two dominant approaches to this line of reasoning: the subjective, which understands the intrinsic value of nature to be found in a subject valuing nature *for nature's own sake* (cf. Callicott 1999, McShane 2007a), and the objective, which understands that value to be found in a specific property or characteristic of nature (cf. Rolston 1988, 1982). In both cases, the world is divided into two kinds of substances, subjects and objects, and their properties. In the subjective sort of view, emphasis is placed upon developing a language that provides reasons to care for and develop appropriate attitudes toward the natural world (McShane 2007a). The objective sort of view typically emphasizes obligations and duties to preserve the property or set of properties that makes nature valuable (Regan 1981).

While there are many variations of the views mentioned above, the point of this chapter is to argue that, from the phenomenological perspective, the approach in general is flawed. Flawed not because of its aims and goals, but because it is asking the wrong question (i.e., on what basis can we establish the value of nature?) in the wrong way. More precisely, asking whether nature possesses an intrinsic value, whether as a result of subjective evaluation or objective properties, approaches the issue from the wrong metaphysical framework, namely the modernist view that the world consists of subjects, objects and their properties. For the phenomenologist, one of the appeals of the phenomenological method is the manner in which the subject-object metaphysics is undermined through the emphasis on intentional relations: If a 'subject' requires an 'object' in order to have an experience and, conversely, an 'object' requires a 'subject' in order to manifest in the manner it does, then the concepts of 'subject' and 'object' cannot be the basic ontological concepts we use to understand the world; there is something prior to the encounter between 'subject' and 'object' that is analysed into those concepts, but it is that prior something, which the phenomenologist simply calls 'experience', that must be understood *without metaphysical preconception*.

This chapter is a phenomenological effort to develop a *phenomenological naturalism*, one that does not separate value from nature, by examining experience without metaphysical preconception. To do so, the chapter is divided into three separate parts. The first is a short examination of why phenomenologists are critical of the naturalism that is typically advanced in environmental philosophy with emphasis on how it gives rise to the question of the value of nature described above. The argument is that the question of how to value nature arises only from the perspective of treating nature as an object with properties. From there, I present an alternative conception of nature and naturalism more amenable to phenomenological analysis in which nature is conceived as a set of relations rather than an object. The consequence of this shift is that meaning and value no longer need to be imposed upon objective nature but inhere in relations, which means that nature itself is suffused with value. By way of conclusion, the final part of the chapter returns to the concerns of environmental ethicists expressed in the original question, how should we live in the natural world? I argue that once meaning is situated within real relations in the world, the normative challenge ethicists face changes from focusing primarily upon preserving specific states of a place to preserving certain meaningful relations. In this way, we are called to focus on humanity's ethical relationship with nature, that is, to inquire into the meanings that are expressed in performing certain actions, to create certain kinds of places rather than others, and to relate to our fellow inhabitants of the Earth with a great deal more compassion than is currently apparent.

MODERNISM AND THE PROBLEM OF VALUE

The phenomenological tradition departs from the criticism of the modernist metaphysical view that the world is composed of two kinds of substances, subjects and objects, and their properties. These two kinds of substance tend to be oppositionally defined, with subjects identified with properties such as active, free, thinking, etc. and objects identified with properties such as passivity, necessity, unthinking, etc. As Val Plumwood has convincingly argued, even contemporary views that are not explicitly dualistic, for example, most forms of materialism, are beholden to this conception of nature because they retain the conception that materiality is divorced from valuing and meaningfulness (Plumwood 1993, ch. 5). In this sense, naturalist philosophies remain dualistic insofar as they allow the dualistic understanding of matter and mind to define what it means to be natural, material. From this metaphysical perspective, then, to be objective is to exist without agency or the capacity to value because those are capabilities of an active subject, while to be subjective is to be contingent and not have any essential relation to the passive objects of the world. Accordingly, if nature, taken as an object, is to possess some value, it must possess this value either in virtue of some property it possesses (objective value) or in virtue of some value it has for a subject (subjective value).

To understand how this metaphysics conditions our understanding of the world, we can consider the following experience: Walking through a southern hardwood forest, I encounter a tree with bright pink, pea-shaped buds, sweeping branches and yet small stature. Amid the various greens and yellows of hickory, black locust, oak and pine, this tree stands out, and I note the graceful extension of the limbs, the texture and mottled coloration of the bark, the spade-like shape of the emerging leaves. Filled with admiration for its distinctness and aesthetic qualities, I attend to the tree for a time before continuing on my way. The modernist explanation of the experience will involve an account of the physical mechanics of perception in my body, the reflectance properties of the various parts of the trees in question and an explanation of the role light plays as a causal agent in allowing my brain to represent the scene and then evaluate it using the various concepts and categories of the mind (e.g., 'green', 'bright', 'spade', 'graceful', etc.) in order to give rise to my experience of admiration. The explanation here is loosely physicalist, with those aspects that do not immediately submit to causal analysis apportioned to the social world, the domain of the subject.[1]

One of the reasons phenomenology is so mysterious to many philosophers is its rejection of such basic and intuitive understanding of the world. What concerns us here, however, are the specific shortcomings of the modernist framework in the context of environmental thought. In this context, Ingrid Stefanovic (1991) has described five consequences of adopting this metaphysics for attempts to define a more environmentally sustainable world. First, there is a tendency to objectify human relation in objects, by which she refers to the process by which we misattribute social significations to nature as if they existed in the object itself. Second, regarding nature as a system of objects facilitates the objectification of the environment into an object of human management and control. Third, our understanding of human habitation is limited to calculated and ordered relations. Fourth, given the other three, we tend to view the implementation of sustainability in terms of imposing an ideal plan upon a passive nature. Lastly, the basic approach to which modernist metaphysics gives rise misdirects our attention away from what should be the normative centre of our ethical enquiry. In her own words, 'if we restrict our definition of sustainability to a balancing of "things", entities, like natural resources, economic concerns and even social, cultural and political factors, we run the risk of taking for granted the fundamental fact that it also and primarily involves the preservation and creation of *quality of life*' (196). This last notion of a quality of life will return in the final section, but, for now, the crucial point is that many of the problems we face in creating a healthier environmental future derive from our very means of conceptualizing the world in terms of dualistic, modernist metaphysics.

If these are the problems of modernity, one must ask how phenomenology proposes an alternative. In its more classical iterations, phenomenology seeks the root of the modernist metaphysics within the lived encounter of an experiencing being and the world. Erazim Kohák describes two shifts that phenomenologists' emphasis on experience engenders in our metaphysical theorizing: (1) from thinking of reality as being 'a set of space-time objects in causal relations to approaching it as a system of interlocking roles' and (2) from thinking of the world as 'an alleged set of space-time objects in causal relations to the *world as experience*' (2003, 22). The world as experience, the lifeworld, does not consist of objects on which meaning and value are projected by a subject existing independently of them. Rather, in the lifeworld we experience ourselves as embodied—with all the limitations, capabilities, dependencies and requirements that entails—among other bodies imbued with meaning and value precisely because of our mutual embodiment.

To see the contrast between this basic approach to the lifeworld and the modernist metaphysics, we can return to the experience of tree appreciation described above. The phenomenologist does not begin with the abstractions of subject and object and try to sort and analyse the experience into these categories. Within the experience, the phenomenologist first finds *an interesting tree*, not a number of sensuous variations judged to be a tree retrospectively through the action of the mind. The tree is not a mere object, however; what it means to be a tree is to possess a limited number of meanings that take on a particular value in light of the history and understanding of the experiencer. In this way, to quote Kohák once again, '*lifeworld*, the world constituted as a meaningful whole by the presence of life, *is neither value free or meaningless*' (25). Meaning is present within the experience through the relation between the experiencer and the experienced, and relative to some larger context against which that meaning emerges (e.g., embodied existence, language, ecological knowledge, etc.). In this way, the tree will always be valuable relative to a sensitive body. Thus, the problem for the phenomenologist is less how an objective property of nature can warrant the attribution of value and more to describe how the various relations present within a given experience express specific meanings and values. In the lifeworld, we confront what Simon P. James has described as our involvement with and inherence in the world: The things that surround us reflect our lived concerns as embodied beings in that world (2009, 21, 34). This position wherein we encounter other corporeal beings that matter to us in various (and frequently ambivalent, contradictory and complex) ways is our basic epistemological situation from which we *could* arrive at a metaphysics of the subject and object by means of abstraction from the experience.

To summarize, from a phenomenological perspective, the modernist metaphysics of subjects and objects (or alternatively of nature and culture) presents somewhat insurmountable problems for environmental philosophy. By artificially separating humanity and our social relations from the larger world of nature, difficulties arise in terms of explaining the existence of value in nature, failing to recognize the role of *interpretation* in both understanding the state of the world and what to do about it, and developing alternative solutions to the problems we face. Phenomenologists propose that the return to the lifeworld sketched above can help to alleviate these stresses, but it remains to be seen what the alternative conception of nature is that it will put in the place of nature as an object.

RELATIONAL NATURE

One of the common criticisms of phenomenology is that it remains a philosophy of the subject, despite its best efforts to remove itself from the context of modernist metaphysics (Latour 1993; Meillassoux 2008). Given the relativity of meaning and value described above, there seems to be some validity to these objections. However, by emphasizing a generic *experiencer* rather than a human being, I am attempting to show that, although human experience must always be the point of departure for a human being, meaning and value obtain within relations in general, not only within relations to human beings. As I will argue, this expansion of the conception of experience from the *life*world to the world of experience can be achieved by focusing on embodied relations, as Maurice Merleau-Ponty did, rather than on conscious relations. If phenomenology is going to present a worthwhile alternative to the modernist metaphysics, it is imperative that it employ a thoroughgoing relational conception of nature based in a more general conception of embodied experience.

To see how we can move from an individual experience to both a general conception of experience and to a relational conception of nature, we can turn our attention back toward the experience with the tree described above. In that experience, the tree is identified through its difference; it stands out against the background of other trees, plants, earth and rock because of its qualities (e.g., its smaller size, its more dramatic shape, the colour of its buds, etc.). But all these qualities are relational: Its size is notable relative to me (about the same) and to the surrounding hickory (much smaller); its colour stands out against the background of greens and yellows, all of which only appear as they do due to the action of light and the structure of my eye; its shape is a product of genetics, the soil, the direction of light and wind, and aleatory events such as interaction with animals, fire, felling of taller, surrounding trees, etc. The experience of the tree includes a multitude of relations that have made it what it is completely independent of myself, but still relative to other beings. The expressed properties of the tree just are its relations to other beings, and in this sense the tree possesses a history, a history of its experiences, of the ways its development has been affected through its relations to others and its environs. These claims are not only true of the tree, but also of all the other inanimate constituents of the environment: the soil produced by the activity of bacteria, beetles, fungus, worms, decaying leaves and perhaps fire; rocks shaped by the wind and rain, boot and burrow; creeks that shift and move their beds in response to the weather and movement of earth. Beings *are* their relations, whether they are conscious of their experience of those relations or not.

This idea that beings are a nexus of relations continually engaged in processes of alteration is common among phenomenologists; what is less common is to situate meaning and value in those relations rather than in the experience of a conscious being. To see the contrast between these two views, we can expand upon the previous experience to include more controversial elements: As I stand appreciating the tree, a feeling of familiarity sets in, and I recognize the tree as one I have seen used in landscaping. For this reason, I being to worry that this tree is an exotic, ornamental tree beginning its 'invasion' of the local ecology. With that thought, I feel conflicted between my admiration of the tree and my discomfort at the possible effects that the tree might have on the local ecology if it successfully reproduces itself. Consulting my field guide, I identify the tree as a native redbud, which ushers in a sense of relief as the previous tension dissolves.

In this example, the value of the tree shifts as it is interpreted differently over time. It would be a mistake to say this process is merely subjective because what is at stake is not the tree's relation to me as a knower, but to the place itself and its other inhabitants. To be 'exotic' in this context refers to the tree's evolutionary status: Did it emerge through its relations to the surrounding forms of life? The local climate, soil and geology? None of this depends on *me*, but it does depend on those other relations. 'Nativeness' is not a property inhering in the tree, however; it is a historical relation to the surroundings. To be native to a place is to exist in a specific kind of relationship to it, hence how we might understand Wes Jackson's (1994) injunction to 'become native' to our places. Similar points can even be made regarding my personal responses to the tree. My admiration shifted to worry because I recognized the possibility of *misinterpretation,* that is, that I was misreading the proper meaning of the landscape and this being's relation to it. As Christian Diehm (2008) describes, identifying trees is a complex process, requiring both learning how to differentiate trees from one another and the different properties of a tree in relation to its seasonally variable environment. I experience conflict because my relations to different *institutional norms* pertaining to aesthetics and ecology lead me to think about the tree in contradictory ways. To settle the dispute within myself, I refer to a third institution, the community of naturalists who have compiled my field guide. Thus, my response is less a decision and more the playing out of the various kinds of relations that form me into the person I am. Though the emphasis in my response is upon my relations, even here it is critical to keep in mind that what is being responded to is the tree's relations to its environs. The value of the tree in this experience, the experience of a human being, occurs in the relations.

Where phenomenology frequently encounters difficulties is when its practitioners rely upon too narrow a conception of experience and thereby present phenomenology as being a relativism to a human experience. In other words, the danger of the phenomenological approach is that the analysis of the human lifeworld is taken to be the only way in which meaning and value come into the world. Lester Embree's (1997) account is a good example of this, wherein nature and its constituents are taken to be imbued with value through their relation to a valuing act of consciousness. Even more sophisticated approaches like that of Charles S. Brown end up leading back into subject-object relations. In an effort to deny the subject-dependence of the ascription of natural value, Brown claims that our perception of value is based upon our biological involvement in ecological networks that operate in a holistic fashion, which in his case means they possess a 'meaningful order of purpose and value' (2003, 13). Thus, the subjective good we perceive is a true or false ascription of value depending on the larger objective good present in the world because, 'there is every reason to believe that this meaningful background of purpose and value has existed long before the human species, and that our specifically human goods only exist within a larger system of good arising from the biotic and prehuman constitution of nature as good' (13).

While we will retain certain of Brown's insights, his view is drawn back into subject-object thinking as a result of continuing to think of goodness as a property that is properly or improperly perceived rather than as a relation.

The danger of subjectivism is what has led David Wood to advocate 'naturalizing' phenomenology by calling attention to different aspects of our natural being that produce us as the beings that we are, for example, how time structures our experience (2003, 213–17). Taken as a relational process rather than a linear flow, time structures our experience in at least four ways: (1) the immediate present conceals the past that produced it as well as the future trajectory of development contained within it, which calls the phenomenologist not only to describe experience, but to engage with it; (2) recognizing the finitude of everything that is calls the phenomenologist to be attentive to the manner in which experience bears within itself a relationship to an entire context of temporal relations; (3) since every being is a finite process, the phenomenologist must attend to the coordination of asynchronous natural processes in the production of the present; and (4) precisely because processes are asynchronous, relations never sink back into a static, holistic system: 'we need a model of the whole [we might say nature] as something that will inevitably escape our model of it' (217, italics removed). Wood's perspective reveals a complexity to temporal processes that transcends our experience of them and calls us to pay closer attention not to how our consciousness constitutes the things it encounters through our relations to them, but to the imbricated relational processes out in the world that constitute natural relations. Even so, regarding these processes as being a part of a system relies upon the 'capacity for demarcating a bounded field' within which relations are possible and therefore valuable and meaningful (223). Doing so, however inevitable, is fraught with dangers because delimiting borders inevitably initiates conflict and frequently violence (Wood 2014). However well grounded, the aforementioned native-exotic distinction serves as an apt example: The distinction itself is grounded in historical, that is, temporal relations and is used to justify the removal or elimination of species found undesirable on its basis.

Wood's understanding of our temporal existence reveals the contingent nature of the wholes that exist, and the fraught manner in which places and environments are individuated, and this contributes something toward a phenomenological conception of nature in which meaning and value are present within relations rather than a projection of a subject. With this in mind, let's return to the basic experience of a body as a nexus of relations to see whether it is possible to assemble the relational conception of nature promised above.

The first step is to begin with a general conception of experience that identifies a being's experience with those relations that have shaped it into the being that it is rather than beginning with a model of human experience.[2] Each body is sensitive to its environment in different ways; these various dimensions through which the body remains open to being affected shape the body's experience of the world. For this reason, we should expand our sense of what phenomenology can do as a result: Since all beings have experiencing bodies, it is, at least in principle, possible to engage with that experience phenomenologically. While all the problems of dealing with another's experience apply, it is at least not absurd to think about a phenomenology of an animal's experience, a plant's or even a rock's. Such inhuman phenomenologies are possible precisely because we share a corporeality with these other bodies, and meaning and value inhere in relations between bodies. By engaging with these inhuman experiences through our shared corporeality, always respectfully, always tentatively, we might develop a better understanding of the kinds of meaningful relations that exist within these very different worlds and the different values that therefore manifest in distinct relations.

If phenomenology turns in this way toward the nonhuman, it is possible to honour the experience of the world without running the dual dangers of (1) modelling the world on human existence, either by conceiving of nature as a subject or as an organism or (2) submitting to a naïve holism that regards ecological systems as somehow inviolable and given in advance rather than a product of a specific evolutionary history. To see why this is so, it is important to understand the shift in the conception of nature that the more general conception of experience engenders. The truth of social constructionism lies in how 'nature' has been defined in many ways over time and has many complex uses to which it has been put. Within the modernist metaphysics, perhaps the most commonsense one is the one already mentioned: Nature is the domain of the object, the necessity of laws that govern the interactions between bodies. Obviously, this sense of nature is not the one to which environmentalists appeal when making arguments for the preservation of nature. After all, nature, on this account, is impossible to destroy since natural laws will exist regardless of what havoc is wrought in an ecological system. This sense of nature, then, gives rise to another more practical sense that can be preserved, namely, the state of an ecological system that would have existed independent of human interaction. This sense of nature will be much more familiar, but note how it is related to the other, more literal sense of nature as an object: Since objects are thought to be authentically what they are independently of the agency of a subject, this sense of nature explicitly employs that dualistic understanding. If we reject the metaphysical conception of nature that underlies this dualism, however, what sense of nature remains?

To answer this question requires us to analyse the interplay of relations across different levels, if you will, within experience. The beings with which we have our everyday dealings are already extremely complex beings. Chairs, cells, other people—all these are composite beings and, as such, we can inquire into the relations that constitute their being. Some of these relations we also individuate as beings—a board, organelles, a community. At the same time, we can look at the larger system of relations to which that being belongs—a dining room set, a pack, a culture—and discuss the aggregate as a being. In our everyday experience, this process of reification of relations into an individuated being is how we make sense of the world. What it reveals is that, though there is only one way to be, namely, to affect other beings and thereby be open to affection, there are as many kinds of beings as there are ways to be affected and affect others. Hence, when speaking of a given phenomenon, it is critical to be mindful of both the kind of being with which one is dealing, that is, the specific assemblage of affects and relations of which the being is capable, and the level one is addressing. Common to all levels, however, is the phenomenon Wood already indicated, the manner in which relations between bodies create spatial and temporal relations, establishing borders in the process. Here is the first relational definition of nature, analogous to the moderns' conception of nature as the great object: *nature is the processes of spatialization and temporalization that occur between bodies*. While this nature can also not be destroyed, it helps us to understand the human role in the environment more clearly. Human presence is just another way of introducing, regularizing and ordering spatial and temporal relations between beings.

To recapitulate, the phenomenologist, due to a commitment to beginning with experience without metaphysical preconception, must abandon thinking of the world in terms of subjects, objects and their properties since these are abstractions from experience rather than something encountered in it. By attending to how bodies interact and individuate themselves, the phenomenologist discovers an extremely general conception of experience in which experience is understood in terms of the ways in which a body can affect other bodies and is open to affection. Within experience so conceived, one encounters myriad bodies, bodies that are differentiated by means of their own relations to one another and their relation to the perceiver via the various dimensions through which the perceiver is open to being affected. These relations can be examined at various levels, which are organized in loose hierarchies (e.g., cells to tissue, tissue to organs, organs to systems), but these relations between levels do not need to be fixed and unchanging. Far from being metaphysically neutral, then, phenomenology discovers in experience a process metaphysics, one in which the world is in a continual process of becoming. From this perspective, nature is simply the name we give to the processes of spatialization and temporalization occurring between bodies. When we speak of nature in the environmental context, we are referring to how these processes operate at the level of places: Nature is the manner in which the bodies within a place establish a specific spatial and temporal order between themselves, always open to outside influence, disruption, the introduction of new relations, etc.

Since value occurs in and through these relations, the world is also permeated with meaning and value. Value is neither a thing that a being possesses independent of relations, nor is it merely the projection of a subject upon an object, which is why the question of intrinsic value is misguided for a phenomenologist. Nature, from this perspective, is a process that includes social relations and so, on this model of experience, there is no metaphysical differentiation between what is typically identified as the natural and the social; both are ways to influence the development of bodies. If all this is the case, then the normative questions environmentalists ask also need to change: Rather than seeking to preserve nature, we should be asking what sort of nature we should collaborate in the production thereof.

FRIENDSHIP AND BUILDING BETTER PLACES

In this final section, I will conclude by saying a few words about how the metaphysical shift described above can influence the ways we think about place building and our relationship, both individual and cultural, with nature. The very idea of a relationship with nature sounds like a dualistic one: Nature exists as something separate from humanity, as an other with which humanity needs to cultivate a relationship. However, it is not necessary to conceive of nature as *absolutely* other in order to speak of having a relation to it if we think about nature as embodied in particular places. From this perspective, it is easier to see how humanity can simultaneously be taken as a part of nature and also needing to develop a relationship to the porously bounded whole. Human activity plays a role in shaping the constituent relations of a given place, and hence we metaphysically belong to it. Approaching the question from a different level, however, we might discuss the relationship humanity has to the place itself in the same way we speak of the relationship specific groups have to larger social institutions. Hence, we can envision improving the relationship that humans have to their places by investigating the quality of relations that certain practices enable or forbid. This approach takes for granted that relations are value laden and embody particular meanings.

One way of considering how to add content to such a view is to ask what meaning we intend our actions to have within the larger biosphere. Neil Evernden (1992) asks this question with his example of the alien ecologist: If an ecologist from another planet were to examine humanity's role within the biosphere, what would the ecologist describe? Would the ecologist liken us to a blight species that overruns and colonizes the ecological systems it encounters at the expense of other species, like a pine budworm? Or would we be perceived more akin to a coral species, providing a home for others in our own biological development? Since we more frequently think about environmental ethics in moral terms, that is, with an eye toward the permissibility of an action, we thereby neglect this inherently ethical dimension of such choices: What kind of species does such a thing? In this way, we acknowledge the inherently narrative dimension to ethical practice.

As I have suggested elsewhere, aspiring to be a friend to nature can be a productive way of envisioning this relationship between humanity, the places it inhabits, and the other species with which it shares those places (e.g., Bannon 2015). Friendship is an interesting model because it combines a number of factors that are, for the most part, excluded from more morally focused analyses. First, its primary focus is the production of a certain kind of relationship rather than the more liberal approaches to ethics of examining individual harms and benefits or the adherence to agreed-upon rules. Doing so shifts the normative focus of our ethical discussions from whether an action embodies impermissible violations of duties to a consideration of whether we are preserving a certain kind of relationship in performing the action. Second, the norms of friendship retain the relational autonomy of nature. In a friendship, we are by definition interacting with the other in ways that preserve the independence of that other. It would be a rather bad friend who attempts to transform their friend into an image of themselves. Third, friendship leaves open the kinds of practices that constitute a productive relationship to a place, which models the evolutionary character of nature itself. We do not desire our friends to stay the same, but help them to continue to change and grow along with us. Lastly, friendships can be fraught at times, but for that very reason provide for means of adjudicating disputes and leaving room for forgiveness.

Although this list is far from exhaustive, one thing to note is how there are no specific principles to follow in making a determination regarding what to do in particular cases. In place of a principle, there is a guiding ethos, an attempt to be a certain kind of person, living in a diverse society, in places that enable the flourishing of all. We will, of course, fail from time to time to embody this ideal of the friend of nature. Instead of viewing these inevitable failures as somehow marking our inability to live productively within the natural world, we can take a more pragmatic stance toward them and try to rebuild and repair what has been damaged or destroyed.

To close, then, a phenomenological approach to the question of what nature leads us to a perspective in which we no longer need to ask the question of whether nature has value or why we ought to value it since all things, in their interactions, are in the process of building meaning and value in the world. The project is, therefore, not to discover some inviolable value that demands respect, but the more difficult political project of determining the value of values: which are to be preserved in which contexts, what is a value in nature hitherto unrecognized, or even which values we currently hold might actually be inimical to friendlier relations to nature. Such a change in focus displaces the environmentalist narrative of humanity destroying nature by our mere presence and allows us to envision a more positive role for our species cooperating with others to produce a better world.

NOTES

1. There are obviously other possible accounts, as many as there are contemporary philosophers' variations on the modernist metaphysical framework. For example, some eliminativists will not admit that anything resists physicalist causal explanation and deny even the experience of a subject as an illusion, while supervenience theorists will seek out the 'bridge' between the causal process and the conscious experience. While I acknowledge that my account glosses over these complexities, the simple point I would make despite the panoply of views is that they all share in common the same metaphysical commitment to explain (or explain away) how bodies, which have the properties of objects, can give rise to putatively subjective experience.
2. I argue this point in much greater detail in Bannon 2011.

Part II

Metaphor, Agency, and the Human Relation to Nature

SIX

Intersubjectivity, the Environment and Moral Failure

Mark Thorsby

Today the environment suffers from a great moral failure. Scientists around the globe are dutifully recording their descriptions of an anthropogenic extinction event while modern life goes on relatively uninterrupted (Kolbert 2014). Simultaneously, massive evidence points to an anthropogenic climate-warming event that will exceed ecological thresholds (CCSP 2009, 97). Both events will have enduringly irreversible results and both are being caused in large measure as a consequence of collective human action. The activities that are contributing to the destabilization of the environment reveal a literal consumption of nature. Unlike ordinary moral problems, these activities are both collective in arrangement and temporally staggered over generations. It is not one but all who bear responsibility. This creates two extremely pressing problems for the efficacy of most moral theory.

First, the assignment of moral responsibility, and consequently that of making moral judgments, becomes extraordinarily difficult because of the diffuse nature of collective harm. No one individual can be held morally responsible for the climate crisis because no one person's action could ever be a sufficient antecedent cause. Problems of collective agency require an understanding of collectivity itself, what phenomenologists call intersubjectivity. Second, collective action confounds most moral evaluations because of an inability to recognize our activity as morally relevant to the environment. Anthropogenic climate change is the result of a multiplicity of actions, taken for differing reasons that in the aggregate are accelerating the accumulation of greenhouse gases into the atmosphere. The individual agents involved do not perceive their daily activity as dangerously harmful because of the varied and temporally staggered nature of collective action. Consider the environmentalist who flies around the country giving papers about climate change—her actions contribute to the very problem her actions are meant to stop. Each of her actions, taken in isolation, are insufficient as causes for climate change, but without the involvement of individuals like her there would be no climate crisis. What is even more disturbing is these activities appear to be accelerating (IPCC 2014) despite a history of public discussion and rational argumentation going back to the 1970s and 1980s (Behringer 2010, 191), if not considerably earlier. There appears to be a breakdown between our documented understanding of the natural world (that clearly outlines the facts of a crisis, especially in its natural scientific sense) and our ability to experience it as a moral crisis. This breakdown, or incapacity, reveals a stunning moral failure and the point of departure for this chapter.

Environmental moral failure begins with the experience of a moral choice. The description of that choice is organized by a requisite cultural world *w*. This cultural world is intersubjectively constituted and acts as a background for the assumptions and reasons that are weighed in practical decision making. Whatever choice is to be made, a reason for action can be given that conforms to some moral principle *p*. Once a decision is made, the practice is implemented. Over time, that practice is habituated in the course of conduct with infrequent or little reevaluation. If the practice involves other agents, then over time these practices are sedimented into an intersubjective harmony of activities. Since the environment is a constituting condition for the world, a worldly (i.e., intersubjective) activity that fulfils a worldly purpose but simultaneously detracts from the constitution of the world becomes contradictory to itself over time. In other words, actions that threaten the existentiality of the world are by default morally contradictory.

The following chapter will be divided into two main sections: (1) the phenomenology of intersubjectivity, and (2) environmental moral failure. In the first section, I explore and define the concept of intersubjectivity within the field of phenomenological philosophy, laying out some of the foundational elements of genetic phenomenology. The second section of the chapter seeks to articulate a conception of intersubjective moral failure that might help us get a conceptual grasp on the moral issues that arise out of collective activity.

THE PHENOMENOLOGY OF INTERSUBJECTIVITY

A phenomenological investigation of intersubjectivity is essential for understanding the structure of the worlds that communities inhabit, build and proliferate. The cultural traditions, assumptions and histories that make up our sense of the world (and our place in it) are intersubjectively constituted; that is, every culture is created out of a combination of mutual interactions and interrelations between the individuals that make up that culture over time. No one person is responsible for any culture, but every culture is composed of individuals living a specific form of life in practice. Intersubjectivity is the phenomenologist's term for thinking about how the intentional experiences of multiple individuals interrelate. Much of classical phenomenology is ultimately devoted to articulating the necessary structures of conscious experience, what Edmund Husserl calls eidetic structures (Husserl 1983). But any discussion of ethics in phenomenology requires a thorough working out of the intersubjective conditions foundational to any community as such. The environmental disruption of climate change and mass extinction, for instance, result from collective activities such as pollution and urbanization. In like fashion, any 'solution' or response to environmental disruption must begin with a clear view as to how communities are formed as collectives. As environmental ethicist Dale Jamieson has argued recently, the types of problems indicative of climate change are such that individual action in isolation is effectively useless and requires a deeper understanding of the world and our relations to it and each other (Jameison 2014). The phenomenological analysis of intersubjectivity yields an ontological description useful for identifying and modelling types of environmental moral failure. Today we are faced with confronting world-constituting problems and this requires the clarification of the intersubjective field that does the constituting.

From the outset, and before considering any problems in ethics, we need to say something about what intersubjectivity means in general. The term itself can refer to a number of potential things. Despite having 'subject' in its name, intersubjectivity should be understood as referring to the space between or wherein subjects *exist* with one another. It is important to understand intersubjectivity in an existential register. By existential, we are saying that the concept refers to a specific form(s) of *being* in which we are with others. The notion of intersubjectivity I employ here is one in which intersubjective relations are *concrete* forms of social activity and consciousness. There is a tendency to view intersubjectivity as a mere extension of subjectivity such that intersubjectivity is treated as a nominal category of description and not a real state of affairs. Nominal intersubjectivity tries to situate the relation among subjects in the mind of the subjects alone where collective action is reduced to a multiplicity of individual subjective acts; it is ultimately a denial of collective agency. By contrast, my thesis is that intersubjectivity is a real feature of the world which is given in materiality. While it is certainly true that individuals have subjective experience—that is, they have their own perspectives—there are real concrete forms of existing intersubjectively that are independent of subjective experience. For instance, consider the example of a 'market economy'. An economy exists as interrelations of exchange between varying individuals. While we can say that an economy is an amalgamation of human interactions, we should not say that is the sum total of individual subjective experiences. Why? For the simple reason that the economy, even the so-called virtual economies, are forms of material activity. Consider the fact that economic data tracks material indicators. An 'economy' exemplifies an intersubjective existence that is independent of individual subjective experiences. In saying that intersubjective relations are real, we do not need to 'point' to some object in the world. Intersubjectivity is given in the repetitious patterns of coordinated movement among subjects. The existence of intersubjectivity is given in space through time and consists in the regulated and meaningful embodiment of the subjects themselves.

Intersubjectivity is fundamentally allergic to forms of solipsistic reasoning in which subjects are seen as epistemically in isolation from one another (Husserl 1998, 12). The topic of intersubjectivity has in recent years proliferated as an important interest among a variety of thinkers, including psychoanalysts, gender theorists (Benjamin 1998) and social theorists (Habermas 1973) but it has its roots in the phenomenological philosophy of Edmund Husserl. Phenomenology seeks to clarify the ontology of what must necessarily be the case such that we experience the world as we do. Intersubjectivity is a constituting condition for meaning and subjectivity itself. The subjective experience for any individual dynamically develops according to pregiven intersubjective parameters and conditions. Husserl goes as far as to say that the 'intrinsically first type of being, that which precedes and bears every worldly objectivity, is transcendental inter/subjectivity' (1998, 38–39). Intersubjectivity is an ontological state of affairs characterized as a harmonious interrelation between subjects. I argue that these relations have a concrete existence in embodied life. An investigation into environmental moral failure needs to begin with an ontological description of what collective action consists of in the first place. Only by first knowing how collective actions relate to individual conscious agents can we properly explain the constitution of arrangements that are literally causing widespread ecological harm.

A primary goal of phenomenology is to uncover the meaning of our experiences; this requires a description of those eidetic and ontological structures necessary for consciousness. There are two types of phenomenological method: (1) static phenomenology and (2) genetic phenomenology. The first form of phenomenology seeks to understand the conditions of meaning for any specific experience as such while the second seeks to understand the generative conditions that enable our experience of meaningful phenomena. Static phenomenology takes the consciousness of the ego as already constituted and whole while genetic phenomenology seeks to articulate the conditions around which a subject develops over time through experience. Genetic phenomenology, by contrast, seeks to uncover and elucidate the laws of genesis that govern the temporal development of consciousness over time. For instance, consider visitors to the Muir Woods, who stand in awe of the great sequoia redwood forest. There are two ways in which we can say visitors have a meaningful experience. On the one hand, they must be immediately conscious of the forest as something there before them; that is, they 'see' the trees as objects before them. But the meaningfulness of the experience depends upon an additional act of reflective consciousness in which the agents grapple with their own identity as juxtaposed to the mighty redwoods they see. A contemplative moment includes both (a) the static moment of seeing the forest with (b) the historicity of one's own developing identity. Genetic phenomenology seeks to articulate the laws of development over time. Edmund Husserl called the genetic analysis of experience 'eidetic reduction' (1960, 72), and it consists of analysing an 'object' as one among all other possibilities for that thing. For instance, I treat this tree here as one among many possibilities for this thing. I can easily imagine this tree as a young sapling or of a different shape, etc. One reduces the content of phenomena to a minimal fixture of essential description that traces the laws and structures that govern the development of that thing. The intersubjective community genetically organizes the identity of moral agents, and the means by which moral agents conceptualize their relationship to nature.

There are two types of genetic development: (1) passive genesis and (2) active genesis. Let us begin with the latter. Active genesis refers to the form of generation in which the ego actively generates its identity (a synthesis of one's built-up memory) through its own subjective processes, such as making moral judgments (Husserl 1960, 77). Acts of practical decision making, logical argumentation and other active subjective processes each contribute to the active genesis of the moral agent whereby the individual constitutes 'new products' for habituation. Anything actively constituted presupposes a deeper principle of passivity (78). One key form of passive generation is the type of habits we employ over time which play a constituting role in terms of the development of moral agency. It is true that the act of making a decision falls within the description of active generation, but once a habit is set and becomes normalized, then it contributes to the passive generation of the moral subject. Passivity is pregiven ahead of active processes, forming the context wherein our decisions play out. The passive genesis of the subject includes more than just the habitual life of a moral agent taken in isolation but also the harmony of habitual relations that surround an individual, the intersubjective. The day-to-day experience of life within a community is one that is both active and passive. In an active sense, I can engage with others in various forms of communication. But the majority of my experience of a community consists in the repetition of silently harmonized movements of the body. Consider how the subtle bodily movements of a person can either put one at ease or in a state of anxiety. This form of perceptual recognition is predicated upon an already established set of normative habitualities. Morality is about the interlocking relationships we have with others (Husserl 1998, 34). The passive generation of the individual decision maker is intersubjectively founded. We must distinguish our experience of intersubjectivity from the ontology of the intersubjective, the difference between inter*subjectivity* and *inter*subjectivity.

Intersubjectivity begins with the body. When I recognize that other moral agents look and act in a manner that is akin to my own, I begin to recognize their actions as intentionally directed like my own. In other words, I begin to 'pair' (Husserl 1960, 112) my reasons and intentions with others. We now know that the neurology of the brain provides fantastic biological mechanisms that make this pairing possible (Zahavi 2014, 197). How well I can pair the intentionality of other agents to the movements of their bodies is an important criterion for determining where on the spectrum of intersubjective familiarity I stand with regard to others in my community. Through language, the intentionality of the other can be paired to a high degree of sophistication. Language operates as the fluidic interface I have with others, a cultural product that is not rooted in the subject alone. The symbols we use to construct our self-identity come from our community and are not subjectively construed. The capacity I have for recognizing the other as like me is what Husserl and Edith Stein call empathy. Stein describes the literal act of empathy as the 'comprehension of an act of feeling' (Stein 1989, 92). Part of the problem for environmental ethics is that our experience of nature is one in which we experience nature as silent. There is no language or intersubjective relation to our experience of the environment. We have an experience of interactivity with nature, but not a meaningfully intersubjective one.

While the experience of intersubjectivity is described in the language of a subject's first personal perspective, the intersubjective field is ontologically distinct. I use the term 'field' here because intersubjectivity taken in terms of its concrete existentiality is a collective harmony of embodied movements over time. The intersubjective field is organized by a common language and grounded in shared embodied practices habitualized within the life of a community. A community can be designated as an intersubjective field where the subjects of the community have a set of shared passivities and activities that are largely synchronized in language and body. Every community has a collective identity that is distinguishable from other communities. Because the community plays an antecedent role in the constitution of an individual's identity, we can say that the community sets the parameters for the meaningfulness of a given moral agency. Many of the norms that are established in communities persist as passive habitualities long after the originating agents or reasons are forgotten. In terms of our problem of environmental moral failure and the problem of collective agency, we can begin to sketch an answer to our original problematic. Collective moral action is intersubjectively organized in forms of both active and passive generation. If certain passive habitualities exist that are deeply sedimented into the life of a community but also harm the environment, then those activities might become morally invisible.

ENVIRONMENTAL MORAL FAILURE

Recent philosophers have begun to look at the problem of moral failure more seriously. Lisa Tessman's *Moral Failure: On the Impossible Demands of Morality* situates moral failure in the space between of an agent's saying, 'I must' and 'I can' (2015, 256). In her view, moral failure is fundamentally tragic insofar as it consists of a normative demand in the face of performance impossibility. Other authors on the topic include Julie Tannenbaum, who argues that moral failure consists ultimately in an inability to live up to obligations (2015, 58–84). She argues that certain challenges in fixing the fulfilment of responsibility can be avoided if we designate a category of 'mere moral failure' to the tool set of categorical distinctions available to ethical judgment. Both Tessman and Tannenbaum situate moral failure as a category related to individual moral agency. But what use can such an account of moral failure have if the type of failure is stretched over a collective group of agents over time? The type of moral failure that the climate change crisis represents, for instance, cannot be easily distinguished by an agent-centric concept. Intersubjectivity requires another conceptualization of moral failure.

For the sake of argument, we can distinguish at least two different things referred to as moral failure. The first is what I call *moral failings*. A failing refers to the inability of a moral agent to fulfil some moral duty. The concept of 'mere moral failing' Tannenbaum describes most closely aligns with this. To fail in the fulfilment of a moral imperative is to fail in one's apparent responsibilities. There are litanies of reasons that people have moral failings—from the corruption of character to bad luck—but in each of these cases moral failure has the structural characteristic of a performative break. Moral failings are not necessarily tragic, however, if by 'tragic' we mean what Whitehead refers to as the 'the solemnity of the remorseless working of things' (1948, 17). Moral failings might be tragic, but not out of necessity. The sense of tragedy is better understood as a sort of paradoxical condition in which a moral failure is the unavoidable result of agents' attempt to fulfil their moral duty. It is this second sense of moral failure that I think is at issue in the case of the environment. My thesis is that environmental moral failure is the collective result of passive habitualities that are simultaneously in conformity with a moral principle. Consider that the moral technician of a coal-fired electricity plant does not engage in her daily activity with intent to harm the environment. Rather, she will likely see her daily activities (the very same activities which are instigating ecological harm) as a form of fidelity, or a commitment of obligations she has to others around her, whether they be her boss, her family or herself. The activities she takes in her natural attitude are world reinforcing despite the paradoxical and deleterious effect on the natural environment as a whole.

A *moral failure* is any case in which a moral theoretic principle (or reason for action) is adopted but hitherto through iterations of activity leads to the contradiction of the original principle. In other words, moral failure refers to those cases in which the successful fulfilment of a moral principle leads to the inevitable contradiction of the same principle. Accordingly, this is a more radical conception of failure than that of a moral failing. Moral failings are unfortunate and moral failures are unavoidable. Even a sound moral theory can be conceptually, or theoretically, non-contradictory and consistent yet still fail when constituted into existing intersubjective practice. This problem is especially salient for environmental ethics. Environmental moral failure corresponds to those cases in which a moral performance satisfies the principle as conceived from the perspective of a cultural world *w* but simultaneously detracts from the constitution of the 'environment'. Industrialism may be destroying our atmosphere, but it has also led to better lives in general for millions of people. This point is perhaps hotly debatable, but infant mortality is down, slavery has been largely extinguished, education is more readily available today than ever before, more people have access to clean drinking water today than at any other time in history, sexuality is no longer a primary factor in the division of labour, and so forth. All of these things are effects of modern industrialism and yet industrial development is threatening the stability of the planet's ecological systems, and by consequence those things as well. The snake eats its own tail.

From the perspective of the individual agent, collective environmental moral failure is conspicuously absent in the field of experience. The natural environment is the material frame for our conception of the world, but how we understand whatever obligations or responsibilities we have to the environment is constituted according to the cultural world we inhabit. It is therefore critical to any environmental ethic that we disentangle and clarify the relation between the environment and the worlds we inhabit. Environmental moral failure begins with the schism between the world we experience and the material environment we inhabit; because one's sense of world is rooted in the intersubjective community, our experience of the environment (in terms of the natural ecology) is preorganized in a nonneutral way.

The 'environment' is the concrete ecological and material field wherein our activities occur and interact. It is a constituting condition for our experience of the world such that without it all intersubjective activity would cease. By contrast, the 'world' is to be understood as the experience of surrounding conditions. For instance, consider a city-dweller's experience of the city. While the concrete infrastructure is akin to an ecological desert, the city-dweller experiences a geography that is alive and meaningful. This concept of world is closely linked with the idea of a cultural paradigm generated through sets of experiences that are constituted collectively over time with sets of dynamic cultural conditions, assumptions and expectations all habituated into lived activity. In the course of lived activity, our sense of the world—what Husserl called the 'natural attitude' (1983, 51)—operates as a background condition, or reference point, for decision making. The environment is the materially alien starting point that comes ahead of the natural attitude and the cultural world. The cultural products of a community have their comprehensibility in terms of their being a part of the world and every world requires an environment. The type of environmental harm we have in cases like climate change or anthropogenic mass extinction are caused by collective forms of action, both active and sedimented passivities, taken for reasons that have a basis in the cultural world of the moral agent(s). In daily experience, the environment recedes as a background condition for the concrete practices that make up intersubjective life. We experience nature as a baseline condition for moral action and not typically as an imperative for moral action.

A telling problem in contemporary environmental ethics is what philosophers call the demarcation problem; it concerns how we are to demarcate intrinsic and instrumental value. Something has value instrumentally if it serves as a means towards some end whereas intrinsic value is based upon an inherent quality of the thing itself that should be respected for its own sake. For the utilitarian, all actions and things are instrumental towards the intrinsic value of happiness (or preference) itself (Singer 2011, 13). In Kantian moral theory, all agents are intrinsically valuable because of their rational capacity (Kant 2002, 83); other beings like animals have an instrumental value because they lack the intrinsic quality. Environmental ethicists have typically argued for the extension of intrinsic value into the natural. But if the natural world is also the material basis and background condition for any community as such, how can the intrinsic and instrumental values be demarcated? Whatever we might hinge intrinsic value on, it will inevitably 'include some things as valuable that we do not wish to include . . . or else we will be forced to exclude some things that we do not wish to exclude' (Samuelsson 2010, 247). Janna Thompson first raised the problem some time ago, and many philosophers such as Lars Samuelsson (2010) and Katie McShane (2007a, 43–61) argue for various solutions. But the demarcation problem is a problem for the very reason that whatever sense one has for either intrinsic or instrumental values, that sense is derived from the intersubjective world. At the level of reflection, values are the names we apply to the categories of objects and activities within our experience or the world. It is circular to apply a label derived from the world to the environment (that is prior to the world) and then to treat that value as itself intrinsic to the environment. Thus, the ethicist who seeks to 'discover' environmental intrinsic value has a false ontological picture of valuing in general. Instead of asking if nature has intrinsic value, we ought rather to ask what sorts of values one's world concretely embodies.

Earlier I discussed the idea that intersubjectivity is based upon a language of pairing with the other. In an important sense, the environment is the context wherein I can recognize the possibility for a pairing of intentional comportment. It is not possible, or least it is difficult to imagine, how one could 'pair' to the environment when it is the means of differentiation. This means that it is not possible for us to experience nature intersubjectively. The natural environment does not have a language with which to voice itself as an agent in which I am in communion. I may certainly have an experience in which I commune with nature, but this experience is not qualitatively distinct from the world that constitutes my experience. Accordingly, I do not think it is possible to situate a phenomenological sense of direct empathy towards the environment. It might be possible analogically to recognize, through the language of science, as it were, that the environment has its own interrelation and niche webs of interaction, but any empathy I have will be indirectly organized as a type of sublimation in which I let my activities with the environment of nature take on the character of intersubjective empathy. I can, as it were, indirectly extend my sense of the world to the environment. The indirect extension of intersubjective empathy towards the environment cannot take the form of recognizing intentional directness in nature (I cannot pair my experience with the natural world that is damaged by anthropogenic climate change), but I can enact concrete forms of practice that reflect an indirect relation. Because intersubjectivity is passively generative, actions that fulfil a moral duty and contribute to the world become sedimented into the life of the community.

Most of our actions are organized by a matrix of habit, tradition and social convention. Intersubjectivity includes the material pattern of movements made between bodies that are pregiven as sedimented habituality. Various movements of the body become embedded and wedded into our perception of the world itself—this is the generative effect of intersubjectivity—such that to question them would involve a questioning of the world. An example of this would be the industrial revolution. It is well known that the industrial revolution in the nineteenth century has been a primary catalyst for the anthropocene (Waters et al. 2014, 9) and much of the ecological strains we see today. Yet when we compare the standard of living after the industrial age with that of the pre-industrial era, we must admit that there has been significant improvement.

Consider the way a well-trained crew of loggers work together over time. Some activities will be habituated or revised more than others, but as conditions change, practices are altered. Intersubjectivity is important here, because those practices which are intersubjectively constituted over generations of time play a generative role for the consciousness of the individual moral agents. For instance, the moral questions that occur to an agent in the course of a decision can become so familiar and repetitious that they begin to look 'natural' such that some practices can become reified. Our experience of the world organizes our activities in the environment. Much of environmental damage is done not because moral agents intend harm, but because either they do not recognize their experience as causing harm or they lack the ability to imagine anything different. As successful practices become habituated and reified, worldly and environmental conditions will change; but the practice continues so long as it performs within the band of fulfiling the original principle p. We can situate three distinct stages of reification or: (1) practices become habitual and repeated without reflection; (2) practices become traditional and normalized as a collective expectation; and finally (3) practices can becomes institutionally enforced. As they become reified over time, practices increasingly become persistent, eventually dogmatic and even imperceptible. Factory farming, for instance, is morally invisible when it becomes a normalized expectation of the way the world just is. The overproduction of corn becomes institutionally enforced by means of economic subsidy. As various industrial practices have become reified to the point of being ideological, it becomes increasingly difficult for a community to even recognize them as morally deficient.

Intersubjectivity exists not in the minds of the subjects, but in terms of the concrete movements of bodies. Most of the time we conceive of values as intellectual ideas that have their reality in the mind of the subjects. We experience values and describe them as things we adhere to, respect, admire, etc. But the intersubjective sense of values cannot be situated as mere dispositions. Instead, the intersubjective field is one mostly characterized by passive habituality and coordinated repetition. Where do values fit under that model? The first thing to recognize is that values are not 'objects' or 'characteristics' of objects at all. They have a concrete manifestation in reality as embodied forms of activity, in regulated movements of the body. An intersubjective value is situated in the regularity of this harmony. A value is derived as a description of the principle that organizes the harmonic movements within a community. In this sense, a value consists in what is done, not in what is desired. The description of an intersubjective value—something valued collectively—is objectively determinable precisely because intersubjectivity is the generative baseline for objectivity as such. A value is a structural principle for lived activity, not the characterization of a normative goal. It is therefore possible for a community to subjectively hold a value of environmental care but to embody a form of life in which their activities reveal a value of environmental consumption unbeknownst to themselves. An agent may have the (worldly) perception of fulfiling a moral principle in their actions but habitually value the reverse. Climate change and species extinction are examples of what moral failure looks like in its most concrete material form. The industrial development of the planet may have had noble goals, but its collective effect may actually stifle the possibility of a noble world. Intersubjective environmental moral failure can help us explain, describe and eventually confront why it might be the case that the very actions that are causing so much environmental harm are conjunctly the very same activities that have resulted in a more peaceful, happier world. Perhaps it is possible to make sense of how and why it might be the case that a perfectly moral person can contribute to the end of the world.

SEVEN

Metaphor and Weather

Thinking Ecologically about Metaphor, Experience, and Climate

Elise Springer

INTRODUCTION: METAPHOR AS AN ECOLOGICAL PHENOMENON

Given a background interest in moral change, I've been thinking for sev–eral years about the role of specific metaphors in our responsiveness to ecological concerns, and with the general role of metaphor in Val Plumwood's "thorough rethink" (2011) and in David Abram's "new way of speaking" (2010). Clearly, ecologically informed projects are better advanced by some specific metaphorical developments than by others. Metaphors saturate the discourse of ecology as well as of the adjacent fields (economics, moral theory, policy-making) that touch on our responsibilities with respect to the non-human world. A critically developed perspective on available and emerging metaphors, such as global warming, greenhouse gases, sustainability, green business, remediation, ecosystem services, etc., is essential. Surely, ecological awareness requires choosing our metaphors carefully, using the right tool for the job.

Increasingly, though, I find it troubling to approach metaphor in this instrumental way. It is a theoretical shortcoming—but also a moral danger—to treat metaphor as a type of cognitive tool with which we human beings organize our *inner* handling of some target phenomenon *out there*. Metaphor itself is an ecologically relational process. If we neglect this fact, we fall back on patterns of thought and practice that exacerbate our propensity for alienation, pollution, and irresponsibility.

Unquestionably, it often seems *useful* to treat metaphor as a tool of understanding and a tool of persuasion. And to debate the *accuracy* of this metaphor is to play tug-of-war on the smoothest of ice. Still, ecological critique invites us to recognize metaphorical phenomena in less domesticated forms.[1] It invites not so much a new theory of what metaphor is but a practice of recognizing and cultivating alternative metaphors — including alternative understandings of metaphor itself.

What is metaphor, once we set aside not only its old decorative status but also its newer function as a tool? It becomes, perhaps, an incoming wave with an undertow, or an electrical current flashing between sky and ground. Metaphor unsettles our cherished distinction between the whorl of life and the flotsam of inert things. Like a wind, it will admit of no clear point of origin; nor does its movement come to any determinate end. Metaphor is a dynamo reshaping the field through which it moves, affording no firm and enduring leverage.

Let us recognize metaphor as many things, but to start: Metaphor is *weather*. One implication will be this: that all but the most local attempts to harness and redirect metaphor are laden with hubris, like efforts to choreograph the global movements of water and gases. Anyone who is wary of technologically systemic solutions to unstable weather should be equally wary of plans to engineer better attitudes through better metaphors, for metaphor is the weather of cognition.

This claim that metaphor is weather, of course, may seem to come from out of the blue (as so much weather does, we should note!) We can begin to recognize this weather by noticing how things unfold with a particular metaphor, selected by a philosopher, for the sake of spurring ecological engagement.

A TROUBLING METAPHORICAL STORM

In his recent book about the struggle to address climate change and avert its most catastrophic possibilities, Stephen Gardiner frames humanity's current predicament as "a perfect moral storm." Gardiner's invitation to perceive climate struggles in this particular new way is purposeful, and comes with an explicit back-story:

> I borrow the phrase "a perfect storm" from Sebastian Junger's book of that name. . . . Junger recounts the true story of the *Andrea Gail*, a fishing vessel caught at sea during the rare convergence of three particularly bad storms, and ultimately destroyed as a result. Given this, I take a perfect storm to involve the unusual intersection of a number of serious, and mutually reinforcing, problems, which creates an unusual and perhaps unprecedented challenge. In my metaphor of the perfect

> moral storm, the three problems (or "storms") [viz, theoretical, political, and moral] are all obstacles to our ability to behave ethically. Like the *Andrea Gail*, we are beset by forces that are likely at least to throw us off course, and may even sink us into the bargain. (Gardiner 2011, 7 bracketed comment added)

This new challenge, it turns out, is not just to our scientific and political capacities, but to our cognitive and emotional habits. Gardiner's metaphor makes no claim to poetic complexity or intriguing novelty. Its aim is strategic: to shift our attention toward systemic problems and to activate the emotions and responses appropriate to an unprecedented storm.

The extended argument of Gardiner's book has many facets: it surveys an array of scientific, economic, and political factors that drive our escalating emission of greenhouse gases into the atmosphere. Beneath these details, however, the philosophical core of the book is an appeal to virtue ethics, somewhat refurbished, as the moral compass with which we ought to face climate change. The "perfect storm" metaphor is pivotal to the persuasive force of this endorsement. Swept up into desperate circumstances, we find only flimsy guidance in rationalist principles, while a fixation on securing optimal outcomes quickly becomes naïve, hopeless, or unmanageable. What we may and must still do, Gardiner argues, is to live a certain kind of life, one that exhibits virtues of character.

Of course the virtue of prudence would recommend steering clear of a bad storm altogether. This is a route that captain Tyne of the *Andrea Gail* failed to follow, despite warnings, in the late days of October 1991. Lacking records of their final hours, we are in a poor position to second-guess the crew's maneuvers in trying to keep the ship, with its six lives, afloat. We do not know how any member of the crew behaved at the end, but we know what would have been admirable: self-discipline, fairness, and reliance on the well-established habits of a good seafarer.

By contrast, Gardiner urges, our own fate in facing climate change is still unsealed. Humanity's best course, having piloted our way into this moral storm, demands extraordinary vigilance at avoiding corruption. Each of us must buckle down in our commitments to justice and other virtues of character; otherwise these moral assets will be easily dislodged and lost. As we rush to secure our holds, it becomes increasingly hard to keep our balance, for the floor beneath us is heaving erratically. Yet if we slack off on the details, lose our moral footing, and succumb to corruption, then we will have given up on integrity and intelligibility altogether. If we let the tragedy play out—this global escalation of noxious exhaust, and the calamities it threatens to usher in—we will have nothing left to say for ourselves.

We thus recognize ourselves, following Gardiner's suggestion, as sailors standing on the deck of a storm-beseiged ship as it begins to pitch and roll. But it is not just the scene within the metaphorical frame that pitches and rolls; the metaphor itself is precarious. Though climate change demands dialogue and political coordination on behalf of both human and non-human interests, the "perfect storm" rhetoric draws on an agonistic stance toward a weather event, conceived as an inscrutable force outside ourselves. Is it wise to frame weather as our foe in this crisis? And over the course of months and years, can we sustain the kind of urgent adrenaline that is summoned by an oncoming storm?

Indeed, what can we make of the *perfection* of this storm against which we are expected to rally? Consider again the storm at sea. At the crux of the *Andrea Gail* encounter, the storm in question had no official name. A hurricane, christened as *Grace*[2] , had merged its power with two other storms. Witnessing the new massive vortex, a meteorologist dubbed it "the perfect storm," thereby gesturing beyond the beautiful into the sublime. When six small lives are juxtaposed against the sublime, how successfully does our allegiance come to rest with them? Are readers moved by the plight of professional seafarers and their imprudent captain, and not at all by reverence for these awesome fronts — neither Christian nor graceful by now — of sideways water?

In the vignette that informs this metaphor, furthermore, every life that hangs in the balance is human. Unlike the dangers of a globally agitated atmosphere, the peril of a sea storm is remarkably peculiar to human beings. No other species ventures out so precariously upon the sea's surface, incapable of retreat by air and quickly ruined by submersion under the surface. Unlike thousands of ships in centuries past, the *Andrea Gail* carried neither livestock nor human prisoners below decks. Just about the only ones directly threatened by that "perfect storm," therefore, were the human seafarers themselves. If Gardiner's metaphor does harness the alarm of readers, will it simultaneously distract us from the fate of non-human lives?)

Further below the surface, we can mark the undercurrent of masculine drama. Our global perfect storm has become, for Gardiner, a test of virtue—*ar ê te*. But *arête* is a moral excellence inextricably associated, in its original Greek context, with manliness and competition.[3] This aspect of his implicit imagery may not have been a conscious strategy on Gardiner's part, but it contributes to the book's effects. After all, those who worry over greenhouse gas emissions have been *emasculated* in American public discourse. Given that effective agency is conventionally figured as essentially masculine, a critic must choose: either move with that current or expend energy against it. In his urgency to motivate readers, Gardiner ends up colluding with the portrayal of moral heroes as 'real men.' There is danger here, he implies—but glory too!—in testing our mettle against an epic nor'easter. But if there is rhetorical appeal of this struggle between man and storm, so too is there great risk that this metaphor will lead readers astray in their confrontation with climate change and with other political problems.

In historical fact, the "perfect storm" of 1991 obliterated the tiny floating colony of humanity caught in its path. Echoing that outcome, Gardiner's subtitle—*The Ethical Tragedy of Climate Change*—foreshadows a bleak and sobering finality. A tragedy is a narrative already consummated, a catharsis-inspiring horror that half-promises that we can leave it at the theater. Can we accept such haunting closure when the forecast includes the feverish convulsion of our planet? Yes, no, yes, no. . . . The book provokes a heaving discomfort—and rightly so—but some of that discomfort belongs to its choice of tropes.

In these ways, despite recognizing a pivotal role for metaphor in moral change, and despite welcoming *A Perfect Storm* for its breadth of information and reflection, I still find myself bracing against the whiplash induced by its metaphors. This distress hardly serves as an indictment against metaphor as such, or in favor of a plain and didactic moral style. On the contrary, it demonstrates both the depth of metaphor and its propensity for turbulence.

MORE METAPHORICAL INSTABILITIES: LARSON'S FEEDBACK METAPHORS

With such concerns in mind, I organized a seminar[4] to reflect critically on the role of metaphor in ecological experience, trying to discover and share promising metaphorical developments while fostering some charitable patience for the difficulties and hazards involved.

Brendan Larson's book, *Metaphors for Environmental Sustainability* (2011), takes this project up explicitly. His case studies highlight the complex life of metaphorical terms coined by experts whose task is to shape and direct public understanding. Trying to communicate complexity in an engaging manner—and often competing for the attention of publishers and journalists as well—researchers and technologists make their choices from within a field of evaluatively charged metaphors. These metaphorical interventions, unlike Gardiner's, are generally intended from the beginning to 'die' as metaphors, to cement themselves into place as ordinary terms.

One of Larson's central chapter-long case studies probes the connotations and long-term implications of military metaphors in the emerging field dubbed "invasion biology." The language of "invasive species" offers a quick cognitive handle for a complex phenomenon. Without such a convenient label, it would be cumbersome to draw attention to the wide variety of cases in which novel forms of mediated dispersal (generally human-mediated, often inadvertent) have enabled the local introduction and rapid proliferation of historically absent species, while long-established local species retreat and dwindle. The concept of "invasion" offers a descriptive summary with vivid compactness, while also serving, prescriptively, to recruit defensive attitudes. If an organism is "invasive," its mere appearance in a new habitat constitutes an attack on sovereign territory. Hence, declaring war against such species is either a "no-brainer" or is simply redundant, since "invasion" implies that war is already underway.

Experts in such research are themselves are quick to insist that the term is only a vivid metaphor. They explain that the ecosystem disturbances at issue here are a dynamic, historical and interactive processes,[5] not reducible to the vicious agenda of an arriving species. Still, their circulation of this metaphor takes advantage of a stockpile of defensive attitudes which are already viscerally meaningful to citizens' groups, policymakers, and backyard gardeners.

Larson's discussion of this case study shows that the process of metaphorical influence is not *straightforward*. It is not simply that listeners' antecedent notion of "invasion" organizes a public interest in eradicating kudzu vines and zebra mussels—two species whose similarity is otherwise highly abstract. Even as the public's attitude toward *invasion* colors their response to incoming species, there is a reciprocal effect as well. The growing discourse on invasive species depends upon attitudes of military nationalism and xenophobia, and in appealing to defensive anxieties about invasion, this metaphorical usage also reinforces these anxieties and modifies their scope.

This phenomenon itself, Larson decides, calls for a metaphor. He nominates several, and ends up proposing "feedback metaphors" as the relevant term for such interaction (Larson 2011, 23).[6] Experts who coin and promote a metaphor such as "invasive species" have particular expectations about how the metaphor wil be received. Like a novice rock band, they may imagine that amplification will channel their signal uniformly outward from the stage. Moving back toward the stage, however, is a selective echo reshaped by what *resonates* with the place and audience. Because of such interactions, the overall sonic effect, within a few cycles of reverberation, takes on a life of its own; in extreme cases, the acoustically filtered echo soon eclipses the articulate clarity of whatever the performers intended to share.

Could ecological biologists have possibly anticipated that their choice of terminology might align them with the political priorities of agencies such as Homeland Security and Immigration and Naturalization Services? Like kudzu— introduced in the southeastern United States to solve a problem, later discovered to have a choking effect on nearly all the life forms around it—a metaphor introduced to solve a problem, no matter how much promise it shows under controlled cultivation, must be released into a different and wider milieu, where it no longer sticks to its assigned task as a tool or as a domesticated specimen.

Reading Larson's book together, my students and I were struck not only by his catalogue of volatile "feedback metaphors," but also by the paucity of satisfactory metaphors in scientific and public discourse. Even the rhetoric offered by Larson along the way seemed to deconstruct itself, while some of the concepts he tries to set aside end up reasserting themselves like cognitive weeds. The rhetoric of "invasive species" in particular—despite Larson's own misgivings—continues to reappear, without distancing gestures, in his own prose.

Despite noting that metaphors are essential in the work of developing theories and communicating about technologies, Larson is clearly torn. Late in the book he urges, in spite of himself, a kind of "metaphoric code—analogous to the Hippocratic oath—according to which scientists refrain from metaphorical appeals while various kinds of democratic deliberation and scientific cross-examination take place. Given the hopes we had brought to the reading of Larson's book, many of us felt a nagging discouragement at such wavering. Hadn't his title (*Metaphors for Environmental Sustainability*) implied that we might catch and bring home, at the end of the day, a few good metaphors? Had not the theoretical core implied that metaphor is *not* a merely optional cognitive step? Yet no single metaphor withstood scrutiny well; straight out of the box, as it were, every metaphor was turning out to be a *mixed* one.

SOME ANTI-ECOLOGICAL IMPLICATIONS OF A STRUCTURAL MODEL OF METAPHOR

Larson's central claim—that metaphor is generally subject to something like feedback dynamics—is an insight worth developing. But because Larson's account of metaphor draws heavily on Conceptual Metaphor Theorists such as Lakoff and Johnson (authors of *Metaphors We Live By*, and many subsequent elaborations), he cannot himself develop that insight very far. Conceptual Metaphor Theory (CMT) acounts for metaphor by building structural models of its architecture, but *feedback* and related phenomena are dynamic processes. The feral unmanageability of even our most deliberate metaphors, I will suggest shortly, thus calls for a broadly phenomenological account that avoids the structural fixations of CMT. These fixations limit its relevance and mislead us about how metaphor happens, and hence about what it would mean to engage responsibly in metaphorical practices.

But to keep this critique in perspective, we should notice at the outset why Conceptual Metaphor Theory would have appealed to Larson, as well as to many other normatively committed intellectuals in recent decades. The first reason is the *cognitive depth* it ascribes to metaphor, not just in language but in experience and practice. In recent decades, many analytic philosophers and linguists had exoticized metaphor as a specialized rhetorical trick hinging on a literal falsehood. Against that backdrop, CMT rightly insists that metaphor is pervasive, cognitively essential, and continuous with whatever might be upheld as literal discourse. Second is the *embodied* dimension and *prescriptive force* taken to be inseparable from metaphorical cognition. (This is not "cognition" in any merely descriptive sense that could be set apart from the ethical.) Metaphors matter, for many CMT researchers, precisely because of their normative implications; they do not think we can address vital matters by simply retreating to an entirely literal plane. The third attraction is the familiarity of the instrumental metaphors in terms of which CMT presents metaphor itself: system, structure, source; target, mappings, frames. As is so often the case, however, the familiarity of a metaphorical motif stands in tension with its capacity to shift our perceptions radically.

The first ecologically-informed concern to be raised about CMT is how it portrays metaphorical influence as unidirectional. To each metaphor corresponds a sentence-like claim, such as "Knowing is seeing" or "Love is a journey." In every example, a "target" concept in need of better understanding presents as a grammatical subject, and the metaphorical function subsumes it under the concrete and familiar concept that stands in the predicate position. Given that degrees of concreteness and abstractness are presented as traits of the phenomena themselves, metaphorical effects should work in just one direction[7] ; but this precludes the possibility that a "source" concept, such as invasion, might be reciprocally affected by its association with a metaphorical target.

More importantly, cognition, as it is portrayed in CMT, uses socially copied tools, but is narrowly individual and brain-bound in its content.[8] Lakoff attends to environmental issues enough to discuss the role metaphor plays in addressing them, but by likening metaphorical frames to firmware installations, his account cannot recognize developments, such as Larson's feedback effect, that manifest only at a social level, and only over time:

> [E]nvironmental frames are the (typically unconscious) conceptual structures that people have in the brain circuitry to understand environmental issues. To understand the real crises one needs the right conceptual structures in one's brain circuitry. Many people have . . . the wrong frames for understanding "the real crisis." . . . What is needed is a constant effort to build up the background frames needed to understand the crisis, while building up neural circuitry to inhibit the wrong frames. (Lakoff, 74)

A third and related troubling feature of the CMT approach is its implicit juxtaposition of purposive human agents and metaphorical structures as tools. Whatever metaphor is, on this view, must be something *by means of which* agents accomplish their projects. We know something by means of metaphor, or we use metaphor to theorize, explain, show, or persuade. Each of these is a tasks whose point seems prior to whatever contribution metaphor might make to completing the task well. Meanwhile, if a metaphor's results ever seem to betray or surprise us, this would not suggest anything like agency on the part of metaphors, only a shortcoming in how we select and handle our own tools. To be a reliable tool, a decent metaphor must be structure-like. A well-built metaphor will not pitch and roll, nor waver, nor propagate in unauthorized ways. The upper part of a structure must not threaten to destabilize the bottom part; a good structure can be expected not to shift too much when we lean on it. On the CMT account, Larson's case studies can illustrate only the shortcomings of the particular metaphors at issue.

Larson's case studies, however, do point to general patterns. All metaphorical phenomena, more or less dramatically, are subject to the feedback effect. Even this admission, of course, might not unsettle anyone's confidence in CMT. Signs of structural failures, no matter how widespread, do not by themselves dissuade us from relying on structural models. We tend to use a paradigm, as Kuhn illustrates, not because all the evidence points in its direction, but because—in the absence of any live alternative paradigm—evidence against it comes across as mere noise. Hence, our next task is to begin animating an alternative habit of cognition.

METAPHORICAL EXPERIENCE AS THE FLUX OF AFFORDANCES

An alternative to structure-oriented understanding is nicely demonstrated by anthropologist Tim Ingold in his discussion of culturally variable patterns of representing animality and life. Asked to draw an animal, it is "second nature" for most Western children to treat the *outline* of the animal as most essential. Thus a successful animal drawing is organized around a closed curve that approximates the visual contours of a single body frozen in time. It does not occur to most of us that an animal might be drawn quite differently, as a line that follows a recognizable trajectory of motion, like the erratic swoops of a swallow's flight, or a constellation of interwoven lines that illustrate the convergence and dispersal of organic material. (Ingold 2011, 69-70)

What kind of animal is a metaphor? Rather than highlighting the boundary between inside and out, one alternative is to foreground the characteristic movements of metaphor. This would be a *phenomenological* account in a broad sense. As in Emerson, Nietzsche, and James, a "phenomenon" need not mean anything more subjective or ineffable than *what goes on, what happens*.

What, then, *goes on* in experience when we say metaphor is afoot? To experience metaphor is to experience a shift, to be confronted with an influx of significance. Whatever we encounter in a metaphorical moment is recognizably continuous with what it was a moment before, and yet it moves us now in some new and different way. For experience is everywhere inflected with what J. J. Gibson calls *affordances*.[9] To perceive something is to recognize how it calls forth some responses and expectations rather that others: each phenomenon we encounter promises and threatens, permits and demands. In metaphorical experience, affordances move right before our eyes. In some sense, nothing has changed; yet a different shade of significance colors the same scene.

Linguistic trappings, it turns out, are neither necessary nor sufficient for metaphor in this experiential sense. For example, many people with an educated respect for biology would not hesitate to count human beings as animals; there is nothing "metaphorical" about our animal status, on their view. Yet experience may or may not provoke them with a kind of shift in awareness that comes from a freshly vivid recognition of the animality of the human being. To have a fresh experience of ourselves *as animals* is to undergo a shift in attention, expectations, curiosities, and concerns.

Val Plumwood describes such a shift following her nearly fatal encounter, in the wilderness, with a hungry crocodile:

> Of course, in some very remote way I know it happened, knew that humans were animals and were sometimes—very rarely—eaten like other animals. . . . Until that moment, I knew that I was food in the same remote, abstract way that I knew I was animal, was mortal. In the moment of truth, abstract knowledge becomes concrete. You gaze with dumb astonishment as your own death, known only as a shadowy, distant stranger, suddenly rises up right before you in terrifying, technicoloured detail and gasp in disbelief that some powerful creature can ignore your special status and try to eat you. (Plumwood 2012, 10-11)

This influx of awareness—disclosure of her human life *as* a source of animal protein—was neither an essentially linguistic realization nor a toll-like structure in her brain. It was an experience produced by the shifting relation between two bodies as one crocodile and its intended human prey wrestled in the open waters of a flowing river. Only such an undomesticated encounter between herself and a predator could so dramatically expose this latent aspect of her otherwise long-familiar bodily existence. Whatever *food* had previously meant to Plumwood—what it is good for, what is involved in seeking it, savoring and consuming it, digesting and excreting it—was freshly spilling and carrying over; it *applied* to her own person, suddenly, with visceral salience. To call this a metaphorical experience is not to discount Plumwood's testimony that this was a revelatory encounter with truth, but rather to notice that the revelation, here, is not of a disembodied proposition; it involves the incorporation of new affordances.

Dangers, we must note, are not figments of any poetic imagination; nor are they peculiar to human life. Danger is a real feature of ecologically-understood relations; so too is opportunity. Hence wherever there is a creature capable of attuning itself to danger, there is a creature who stands open to metaphorical experience, to the double-take through which one phenomenon discloses itself as another.

A verbal expression, such as "people are always potentially food, in reality" might bookmark such an experience, or even prompt an episode of imaginative reckoning among Plumwood's readers. But to undergo a metaphorical shift in perception differs radically from believing any assertion that pretends to distill its point. Such an experience differs, too, from the experience of wielding a tool so as to further some cognitive purpose at hand. Rather like weather, it *happens*.

One other observation about metaphorical language may help consolidate this distinction between metaphorical experience and metaphorical sentences. For if a metaphorical trend in experience were to run its course in consummate fashion, we would be left unmoved by the sentences that might once have been thought to give them voice. For example, the idea of "sermons in stones" to which Duke Senior bears witness in Shakespeare's *As You Like It* can hold our attention only so long as we take revelation to belong primarily to ecclesiastical voices. It is only when reverence for a "mother" still carries more weight than reverence for "nature" that an appeal to "mother nature" can come across with a distinctive imperative charge. It is only for those who still implicitly treat animals as *better* examples of agency than plants or rivers are that "animism" is a tempting way to capture what's distinctive about so-called "animist" ways of life. (Ingold 2011, 67-68)

A NIETZSCHEAN FLOW OF METAPHOR AND METONYMY

I have argued that the "feedback metaphors" Larson discloses require moving beyond CMT, and I have also drawn attention to something like the experiential *field* of more and less highly saturated affordances within which metaphor moves. But it is Nietzsche's reflections on metaphor that help us understand the dynamics of metaphorical turbulence: whiplash, feedback, undertow. These currents of meaning are neither under our control or entirely outside us. They move through us and weather us;[10] and yet it is also largely through them that we are able to move.

As usual, Nietzsche's touchstone, in telling us about metaphor, is philological: our concept of metaphor draws on the notion of *carrying-over* or *carrying-beyond*. (The Greek *meta-* we recognize in metastasis combines with the Greek *pherain*, to bear or carry, that lives on as the *-fer* in transfer). Such carrying over of meaning is clearly a process, not a structure, and on Nietzsche's account, it is a process that is not just common but constitutive of thought, perception, and life itself. By subsuming all representation under the name "metaphor," Nietzsche invites us to dismiss the fantasy of a representation that bears disinterested witness (telling the truth, the whole truth, nothing but the truth).

The distinctive contribution Nietzsche makes here, however, comes is his recognition of two complementary currents of movement within metaphorical phenomena. Larry Hinman summarizes the crucial aspect of Nietzsche's account this way:

> If we think of metaphor as a "carrying over" from one sphere to the other, we see metaphor permeating the process of knowing. Starting from the immediate data of sensation (the nerve stimulus, according to Nietzsche), there is a "carrying over" in two directions. On the one hand, a metonymical carrying over from those data to an imagined external cause of them, a "thing-in-itself" which is the imputed cause of our sensations of it; on the other hand, a metaphorical carrying over of those nerve stimuli into images. The process continues in both directions. (Hinman 1982, 191)

The contrast between these directions might be illustrated within a game of "telephone." We would be witnesses to metaphor if we were privy to each step along the way, each variation on the utterance as it moves along the line of players. Each iteration highlights some aspect of what came before, while dropping others as if they were irrelevant noise. Metaphor, in this Nietzschean sense (prior to good and evil, prior to questions of legitimacy), simply *happens* as cues propagate forward while mutating in their myriad ways.

Meanwhile, metonymy (in Nietzsche's distinctive sense) flows backward along the same chain, whenever a downstream signal is taken as a re-presentation or sign of the upstream phenomena from which it issued. Metonymy is thus the back-propagation of metaphor's innocent metabolic mischief. The garbled result at the end of the 'telephone' line is funny precisely because we treat it as a window on the initial upstream utterance, even while knowing it must have been dramatically different.

With this Nietzschean pattern in mind, let's consider Larson's case of "invasive species" metaphors again. Metaphor unfolds whenever someone encounters a patch of kudzu, and takes it up *as* an invasion. Metonym, meanwhile, brings an inverse current of movement to the same association: this downstream effect of metaphor (the activation of a defensive response) is read backward as if it *informed us* about the nature of the plant encountered.

Nowhere is the undertow of metonymy more evident than in the highly charged and hyperbolic language of praise and insult. For as much as an insult targets a person by means of an epithet, it is the epithet that becomes increasingly charged with the backwash of so many associations. An insult may not convince a child that her brother Billy *is a girl,* but it easily suggests to her that being taken for a girl is the kind of fate that Billy ought to worry about. And it is in this very way that the metaphorical phenomena of "invasive species" may or may not convince us that plants *actually* invade, but they do serve to increase our anxiety that invasion may lurk where we least expect it, its possibility demands our constant vigilance. *After* invasion biology has made its intervention, the public is alert for invasions of a new kind: invasions entirely *lacking* in armaments, commands, and strategic coordination. This possibility—that in *invaders* can be quiet, stealthy, superficially peaceable, and indeed *beautiful*—would not have gained much traction prior to invasion biology. Now, however, any influx of immigrants may be perceived as invasive! And whenever we take this *as* that, metonym masks our own agency in such uptake; whatever we *take* something to be appears to have been given in experience. In such a way, just as invasiveness provides a concept with which to understand plants differently, but *plants* provide a concept with which to understand what invasiveness can be like.

LIVING WISELY AMID WEATHER

What then can we say about the conditions under which metaphorical currents become most forceful, most likely to sweep unbidden through experience? While a structuralist account foregrounds the cognitive purposes that create an interest in metaphorical tools—a vacuum of practically salient meaning—an ecologically integrated account will attend, equally, to the overflow and surplus of significance that seeps, washes, or rushes inward in metaphorical experience.

Phenomena are more or less charged—pressurized—by what we might call a *meta*-affordance: the demand that we be on the lookout for such things even in advance of their full-fledged manifestation. We say danger *lurks* or opportunity *hides* because these are affordances that demand or encourage vigilance and ready recognition. To be *dangerous* is to be the kind of phenomenon that bleeds into its neighbors, carrying over readily through association. It is for this reason that "war" (for example) serves as the "source" domain of so many metaphorical connections. It is *not* because war is more concrete and familiar to us than things, like argument, that come under its metaphorical sway. Thankfully, after all, many lives include little familiarity with the concrete experiences of paradigmatic war; yet those people are none the less riveted by metaphors of war. Concepts such as war, so long as they are saturated and charged with urgent affordances, press outward to find new metaphorical relevance.[11]

Correspondingly, an ecological critique of metaphor asks more than whether this or that target of understanding (climate change, introduced species, etc.) is adequately illuminated by a particular metaphorical model on offer. It also inquires cautiously into how "source" phenomena become reservoirs of metaphorical charge and achieve their surplus of power. Our stance toward such phenomena—storms, invasions, stigma, heroic victories—is not a static background. It is also reshaped by the metaphorical process. Insofar as the implications of "invasion" (for example) are salient and highly pressurized in our social and situated experience, we can expect them to crop up in surprising places, to seep into our experience, and to spill out again through our responses.

An ecological critique of metaphor allows that metaphor circulates in a social world that is as vast and variegated as the surfaces of the earth. To attempt to block or indict the metaphorical trends around us is to tilt not just at windmills, but at the wind. What we can do, however, is to practice attending to emergent metaphorical possibilities, resisting the temptation to nominate *one* metaphor as the best tool for handling a given concern. A milieu of alternative and incommensurable metaphors keeps the contingency of metaphor near the forefront of awareness.

Wise practice, as we live amidst metaphorical weather, does not revolve around the aspiration for mastery in our deployment of metaphorical tools; nor does it require a set of criteria by which to distinguish acceptable metaphors from unacceptable ones. Metaphors, no doubt, can do great damage, and by listening to which way the wind blows, we may hope to foresee this damage here and there, and to create local diversions or shelters. Beyond that, our agency must aim less toward ambitions to make the weather come out right, in its details, and more toward recognizing that metaphor, like weather, resists domestication.

Neimanis and Walker suggest that we "understand climate change and human bodies as partaking in a common space, a conjoined time, a mutual worlding that we call *weathering*." (Neimanis and Walker 2014, 560) If metaphor is change in the flux of meanings moving around and through us, then we must neither distance ourselves from metaphor nor aim for mastery. Rather than seeking to select and adopt "sustainable" metaphors—metaphors with which to feel comfortable and satisfied—we can weather the flux of metaphors, daring to let their lives intertwine with our own.

WORKS CITED

Abram, David, 2010. *Becoming Animal: An Earthly Cosmology* (New York: Random House)

Gardiner, Stephen M, 2011. *A Perfect Moral Storm: The Ethical Tragedy of Climate Change.* (New York: Oxford)

Gibson, James J, 1979. *The Ecological Approach to Visual Perception.* (Hillsdale, NJ: Lawrence Erlbaum)

Gschwandtner, Christina M, 2014. "Might Nature Be Interpreted as a Saturated Phenomenon?" in *Interpreting Nature: The Eerging Field of Environmental Hermeneutics* (Fordham), 82-102.

Heft, Harry, 2003. "Affordances, Dynamic Experience, and the Challenge of Reification" in *Ecological Psychology* 15:2.

Hinman, Lawrence M, 1982. "Nietzsche, Metaphor, and Truth" in *Philosophy and Phenomenological Research* 43:2, 179-199.

Ingold, Tim, 2011. *Being Alive* (New York: Routledge).

Kövecses, Zoltán, 2010. *Metaphor: A Practical Introduction* (Oxford University Press)

Lakoff, George, 2010. "Why it Matters How We Frame the Environment" *Environmental Communication* 4:1, March 2010.

Lakoff, G. and Johnson, M. 1980. *Metaphors We Live By* (University of Chicago, 1980)

Landau, Robinson, and Maier, 2013. *The Power of Metaphor: Examining its Influence on Social Life* (American Psychological Association).

Larson, Brendan, 2011. *Metaphors for Environmental Sustainability: Redefining our Relationship with Nature* (New Haven: Yale University Press)

Lockwood, Hoopes and Marchetti, 2013. *Invasion Ecology* (Second Edition, Wiley Blackwell)

Neimanis, Astrida and Rachel Loewen Walker, "Weather: Climate Change and the 'Thick Time' of Tarnscorporeality.' *Hypatia* 29:3 (Summer 2014) 558–575.

Plumwood, Val, 2011. "Nature in the Active Voice" in Irwin, ed., Climate Change and Philosophy: Transformational Possibilities (Continuum Studies in Philosophy) (London: Continuum) p. 32-47.

Plumwood, Val, 2012. *The Eye of the Crocodile* (Canberra: ANU E Press).

NOTES

1. This metaphor of the "undomesticated" is pivotal to the critical feminist ecology of Stacy Alaimo in *Undomesticated Ground: Recasting Nature as Feminist Space* (Cornell, 2000), though my use of this idea here does not originate in or reflect substantively on her account.

2. The use of exclusively female names for tropical storms had officially been abandoned by the USA, New Zealand and Australia around 1975.

3. The ship closest to the *Andrea Gail* was its sister fishing vessel the *Hannah Boden*. Its captain was — and still is in 2015 — Linda Greenlaw, a self-described lifelong tomboy who proves (as in *probes*) the rule of this gendered association. Hers was the last frantic radio exchange with the men on board the doomed ship.

4. The seminar, "Ethics and Fluency: Metaphors in Moral Cognition" was held at Wesleyan University in 2012 and 2014. Some of the reflections of this paper were shaped in conversation with student participants, to whom I am grateful. Particular mention is owed to Bennet Kirschner, Neo Sora, Hannah Vogel, Ho-Jae Cha, Evan Keyes, Dylan Awalt-Conley, and Seti O'Brien.

5. For example, see Lockwood, Hoopes and Marchetti 2013, 11-13, which specifically addresses Larson's concerns on the troubling connotations of "invasion" in this field.

6. Larson speaks of "feedback" without any explicit discussion of its metaphorical roots; what follows is my own reconstruction of its metaphorical resonance, not his.

7. An articulation of the "unidirectionality" thesis is explicit in the most widely accepted textbook treatment of CMT. (Kövecses 2010)

8. The CMT tradition has spawned variants, and not all of them are equally addressed by this cluster of criticisms. In particular, many critically inflected amendments to CMT can be found in Landau, Robinson, and Maier, 2013).

9. For Gibson's original definition, see Gibson 1979. For a more philosophical and socially inflected discussion, see Heft 2003.

10. Astrida Neimanis and Rachel Loewen Walker, "*Weathering*: Climate Change and the 'Thick Time' of Transcorporeality." *Hypatia* 29:3 (Summer 2014) 558–575.

11. The account here, despite its naturalistic emphasis, bears some evident affinity to Jean-Luc Marion's discussion of the *saturated* phenomenon. See Gschwandtner 2014.

EIGHT

Ecofeminism, Ecophenomenology and the Metaphorics of Nature's Agency

Tim Christion

> Let us beware of thinking the world is a living being. . . . Let us likewise beware of believing the universe is a machine.
> —Nietzsche

The idea of nature's agency finds wide expression today from ecocriticism to the new materialisms. In my view, however, some of the most compelling expressions are found in ecofeminism and ecophenomenology. Yet, paradoxically, they also offer some of the most compelling reasons to question this idea.

Environmental philosophers have implicated the metaphysical concept of nature in a surprising number of philosophical and political problems. Indeed, the project of defining nature's agency is an expression of this. But the concept of *agency* in 'nature's agency' hasn't attracted the same depth of focus. Cross-pollinating insights largely distinctive to ecofeminism and ecophenomenology, I argue that working through the concept of nature is doomed to fail unless accompanied by an equally ambitious investigation of its relationship to agency.

Specifically, I draw on ecofeminists Val Plumwood and Carolyn Merchant as innovative and compelling defenders of nature's agency, especially as this figures into their larger project of articulating root metaphors capable of capturing a mutual relationship between human and nonhuman agency. Plumwood's model of 'dialogue' stresses the dynamics of communication to express social relations and socioecological relations nondualistically. Merchant's metaphor of 'partnership' is similarly mutualistic, but emphasizes an egalitarian sensibility.

Despite its virtues, however, there are philosophical and political reasons to question a reliance on the concept of agency to articulate these metaphors. Philosophically, agency appears ill-suited to mediate relationships mutualistically. In dialogical terms, for instance, agency captures the 'voice of change', but says very little about what it is to *listen* for it. A model resting on mutual agency thus implies two voices talking past each other—two monologues, not dialogue.

Consistently, in fact, there's a pronounced emphasis on *nature's* agency, not mutual agency. The reason, no doubt, is to invite humans to 'listen to nature's voice' given anthropocentric tendencies to 'dominate the conversation', so to speak. Hence, the project of defending nature's agency ultimately tends to express an *ethos of respect* that invites us to stop and humbly listen, as it were, before the biosphere becomes so humanized that we hear nothing but our own voices.

There are also political reasons to worry about expressions of nature's agency as a strategy to restrict human agency. In the context of socioecological relations, projects expressing an ethos of respect rightly seek to decentre anthropocentric justifications for human dominion over nature. In the context of *social* relations, however, this project risks neglecting marginalized social groups—those already rendered voiceless by the same institutional structures of power that mute nature. But unless privileged interests surprise us by respecting nature's agency at the expense of their own, the disenfranchised should be empowered to politically engage a world infused with power relations. The question here concerns whether a one-sided emphasis on nature's agency over human agency is conducive to the kind of grassroots politics needed to address pressing issues like climate change.

Ultimately, I argue for an 'ethical ontology of dialogical partnership' premised on the mutualistic phenomenon of *response* in lieu of agency to comprehensively mediate social and socioecological relations. Drawing on Maurice Merleau-Ponty, I submit on philosophical grounds that an ecophenomenology of response expressing what David Wood calls the 'middle ground of relationality' stands to mutually intertwine the basic categories of Western thought polarized by the metaphysical tradition (2003, 213). On political grounds, I argue that, fortified with ecofeminist analyses of power, this would also bring the ethos of respect into relation with an *ethos of empowerment*. If the former invites all partners in dialogue to openly and respectfully listen, the latter encourages them to speak truth to power and take responsibility for things that matter.

THE METAPHORICS OF NATURE AND AGENCY

Rosemary Radford Ruether argued in 1975 that the oppression of women and the ecological crisis were both rooted in industrial society's 'fundamental model of relationships', which 'continues to be one of domination'. Once this becomes recognized, she adds, the door is open to unite forces and 'envision a radical reshaping of . . . basic socioeconomic relations' (quoted in Warren 2001, 253). The mutualistic 'model of relationships' Plumwood and Merchant advance is very much in this spirit. Their projects overlap in two respects. First, both are sensitive to the pernicious consequences of dualistically polarized conceptions of relationality stemming from Western metaphysics. Second, each recognizes that the cultural transmission of the metaphysical tradition occurs largely in metaphorical form.

Plumwood argues that the dominant logic of modernity is organized by a 'set of interrelated and mutually reinforcing dualisms which permeate western culture' (1993, 42). In particular, the dualism between 'nature' and 'reason' forms 'a fault line which runs through its entire conceptual system' (ibid.). Some of the most basic categories of Western thought are polarized along this fault line depending on what counts as 'rational' or 'natural'. For example, 'rational humans' are privileged or valued over 'natural animals'. But this same logic applies equally to social relations to the extent that some humans are deemed more 'human' (rational) than others. Expressions of this are myriad: the autonomous husband versus the emotional but nurturing wife, 'naturally suited' for domesticity and dependent on the 'head of the household' to make decisions; the manifest destiny of 'civilization' to conquer wilderness and rationally order the world versus 'wild savages' living instinctively in a chaotic state of nature; the social 'progress' or 'development' enabled by scientific, technological and economic control over nature versus 'backwards' or 'undeveloped' societies ensconced in irrational traditions binding them to the land.

'To be defined as "nature,"' Plumwood explains, 'is to be defined as passive, as non-agent' (1993, 4). Reason, by contrast, is an agent for properly ordering the world, ostensibly to the benefit of all. Hence, the naturally 'irrational' depend on the 'rational', but not vice-versa. Ultimately, therefore, the interlocking set of dualisms structuring this logic amounts to a 'monological' model of relationality. Power flows in one direction: from the autonomous voice of reason to the naturally dependent other expected to listen.

Western culture generally has no problem consistently conceptualizing human agency as a nature-shaping force. Equally conceivable are situations where agency is trumped by natural forces beyond all control. Yet, there appears to be a profound difficulty conceptualizing a mutually conditioned *and* conditioning human relationship to nature without becoming ensnared by dualisms that pull logic to one side or the other.

Debates over the significance of 'wilderness' expose this logic with unusual clarity because nothing less is at stake here than the meaning of pure nature. The debate largely pivots on whether environmentalism should emphasize preserving nature from human agency by encouraging a respect for 'self-willed land', or focus instead on proactively working towards a sustainable society. Accordingly, the conflict is between a 'hands-off' view that privileges nature's agency over human agency and a 'hands-on' view that does the opposite. Reed Noss exemplifies the former: 'Wilderness represents self-imposed restraint in a society that generally seeks to dominate and control all of nature' (1998, 410). J. Baird Callicott and William Cronon, however, argue that a viable environmentalism should wisely *engage*, not limit, human agency. They call for an ethic of social responsibility premised on proactively developing a sustainable relationship to the land.

Since Callicott and Cronon's critique, environmental philosophers have dedicated themselves to nondualistic alternatives to wilderness. According to Plumwood, however, they merely reproduce dualism in different form. Working with only two terms, usually 'culture' and 'nature,' one of three logical options are possible: Keep the terms 'hyper-separated' by emphasizing differences over continuities, as in wilderness dualism, or else 'incorporate' one term into its opposite by privileging identity (unity) over difference. Callicott's 'naturalistic reduction', for instance, incorporates culture into nature, minimizing what is uniquely human. Cronon's 'cultural reduction', by contrast, reduces nature to a cultural construction, neglecting nature's distinctive presence (see Plumwood 1998, 671–76).

In each case, the polarizing logic basic to dualism remains: One term is privileged over the other as an independent reality and agency all its own. Overcoming dualism, therefore, involves articulating *mutual* relations that express identity *and* difference. If, as Plumwood argues, the metaphysical 'fault line' dividing nature from agency amounts to 'monological' relations, 'dialogue' dissolves this dualistic logic at its foundation by recognizing agency in humans and nature both.

Merchant's metaphor of 'partnership' is similarly motivated. As Donna Haraway acknowledged in 1988, ecofeminists have been especially vocal proponents of nature's agency, and this is arguably credited to Merchant's groundbreaking *The Death of Nature* (Haraway 1988, 593). Today's ecological crisis, she argues, is traceable to a dramatic change of consciousness following the Scientific Revolution. Facilitated by a cultural shift in 'root metaphors', this involved transitioning from a vision of nature as a living 'organism' whose integrity and agency was to be respected to nature as a lifeless 'machine' exploited at will.

Root metaphors conceptually model 'a single cultural reality [or] world view' articulating the basic relation between self, society and nature (Merchant 1980, xxii). Premodern conceptions of nature as an organic unity, for instance, assume an *internal* relation between self, society and nature. Like the organs in a body, wholes take priority over parts. Individuals are thus subordinated to society, which is in turn subordinated to nature. In sharp contrast, the machine metaphor expresses a philosophy of *external* relations that privileges parts over wholes. For Descartes, nature is an assemblage of passive atoms externally ordered by mathematical laws transparent to human reason. Scientific methods revealing these laws can thus be applied technologically to enable human control over nature. Applying this metaphor to society, Hobbes imagined atomized (egocentric) individuals in an anarchic state of nature where social cooperation had to be externally coerced by a governing power enforcing law and order. Individuals are therefore essentially independent of others in society, just as humans are from nature.

Western culture today, Merchant explains, embodies an uneasy juxtaposition of these conflicting metaphors. Commensurate with Plumwood, they articulate a mutually exclusive relation between 'nature' and 'agency.' While the organism metaphor offers a holistic view of nature's agency at the expense of human agency, the machine metaphor (in Cartesian guise) suggests an exclusive view of human agency at the expense of nature's agency.

The 'death of nature', then, is a metaphorical 'death', not unlike the 'death of God' proclaimed by Nietzsche. But, also like Nietzsche, metaphor is far more than linguistic flourish. Indeed, as George Lakoff and Mark Johnson argue, metaphors make our 'subjective lives' possible by conceptually structuring our 'everyday metaphysics'—our sense of what is real (1999, 17).

To the extent that this metaphorical metaphysics—'metaphorics', let's say—constitutes the background of thought and experience, the metaphors of *nature* analysed by Merchant are especially important. After all, nature might be the most metaphysically loaded word we have. In this light, Plumwood and Merchant's focus on the mutual metaphors of dialogue and partnership seem promising as a response to what we might term the 'metaphorics of nature and agency'—the dualistic 'model of relationships' basic to Western sensibilities.

QUESTIONING NATURE'S AGENCY

If the metaphorics of nature and agency constitute a 'fault line' organizing a network of dualisms spanning social and socioecological relations, a nondualistic conception of nature's agency would strike at the heart of this logic. Philosophically, however, extending agency to nature risks reinstating dualism, despite Plumwood's intentions. Politically, moreover, I worry about expressions of nature's agency geared to restricting human agency.

From a phenomenological perspective, Plumwood's defence of nature's agency reveals problems with this concept generally. For Plumwood, the first two steps in the development of dualism (credited to

Plato) involve establishing rationality as the normative ideal of human identity, and then defining it in opposition to nature. The third, accomplished by Descartes, constructs nature as mindless, thereby removing any vestiges of continuity between the human and nonhuman (Plumwood 1993, 107). Plumwood's position of 'intentional panpsychism' is thus a countermeasure meant to sensitize us to nonhuman agency by 'extend[ing] our recognition of mind-like qualities much more widely into the world' (2002, 176). Locating intentional agency in nature itself thus offers a 'rapprochement and intermingling of these orders hyper-separated in Cartesian thought' (2002, 179).

Plumwood's proposal, therefore, rests on 'extending' intentionality, once the privileged attribute of human beings, to the nature pole. Yet, without a third term, one is forced—like so many philosophers since Descartes—into futile attempts at 'bridge building'. Indeed, Plumwood sees her position as a part of a larger effort to 'rebuild the severed bridge of mind-body continuity from both ends' (2002, 179). On one side, there are philosophers revealing the 'embodied mind'; on the other, those like her are disclosing 'mind-like qualities' in nature.

In addition to the problem of conceptualizing nature as an agent directed toward its own ends (see Bannon 2009), I worry that intentional panpsychism, and nature's agency generally, remains tied to the same two-term logic plaguing the wilderness debate. If intentional agency suggests a source of activity that changes circumstances, as opposed to being passively swept up by them, this concept relies on a problematic activity/passivity divide. Like the mind/body pair, this more abstract dualism is centrally implicated in the interlocking dualisms structuring monological relationality. Plumwood herself argues that addressing specific dualisms such as man/woman and civilization/wilderness requires reworking the general dualisms undergirding them, like culture/nature and mind/body. Yet, as phenomenologists demonstrate, rearticulating the latter further involves reworking even more abstract dualisms like activity/passivity and interiority/exteriority.

Interestingly, Plumwood invokes Franz Brentano, remarking: 'intentionality has been widely seen by phenomenologists as the most important mark of the mental' (2002, 131). Indeed, it's instructive that the philosophical context leading to phenomenology is similar to the one situating Plumwood's project. At issue is the perplexing post-Cartesian question of how the mind makes contact with the external world. Brentano's proposal hinged on defining mind as *intentionally directed* to things. But he still struggled to explain how it could intend 'real' things as well as 'ideal' things like dreams and hallucinations. Working with only two terms, the intentional subject and the intended object, he waffled between realism and idealism.

It was precisely in response to Brentano's dilemma that Edmund Husserl made his decisive breakthrough by introducing a third term—the 'noema'—to *mediate* subject/object relations. Real objects, like 'that comet in the sky', exist, and there are intentional subjects conscious of it. But what's important is the general (noematic) *meaning* of 'comets' as such that determines *how* a subject can be intentionally directed to, or make sense of, that comet in particular. The same comet, after all, that appears to many medieval Europeans as an omen of disaster and now appeared to many Europeans as a chunk of ice and rock orbiting the sun. Hence, the third term of mediation is intersubjectively *normative* in ways that implicate subject and object but is reducible to neither. Although not a phenomenologist, if Plumwood's philosophy of intentionality was informed by a mediating term, like Husserl but not Brentano, 'bridge building' might have seemed superfluous.

Furthermore, the very concept of intentionality is fraught with problems and complexities. Husserl came to recognize that 'intentional acts' are thoroughly infused with a kind of passive intentionality. That is, expressions of intentionality are so mediated by historical, social and embodied structures of meaning that the very distinction between active and passive is ambigious. For Martin Heidegger and Maurice Merleau-Ponty, intentionality is simply the surface phenomenon of a much deeper ontological structure. Whether employed to build bridges or signify agency, therefore, applying intentionality to nature is philosophically questionable.

Yet there's another way to articulate the nature's agency project commensurate with phenomenology. If, for Merchant and Plumwood, nature *itself* is deemed too complex and chaotic to predict and control, the complementary version achieves the same conclusion by stressing just how thoroughly mediated *human* intentionality is. Some ecophenomenologists thus put forward a Heraclitian nature that 'loves to hide'—one that decisively exceeds our intentional grasp and disrupts our lifeworld expectations. In this vein, David Abram's animistic reading of Merleau-Ponty focuses on 'the intelligence that lurks in nonhuman nature . . . that temporarily shatters habitual ways of seeing and feeling, leaving one open to a world all alive, awake, and aware' (Abram 1996, 19). Intelligence here is defined as 'a property of the earth: we are in it, of it, immersed in its depths' (262). Similarly, Bruce Foltz offers a Heideggerian phenomenology of 'radical givenness' that 'enchants and animates' nature (2004, 340). Accordingly, the job of environmental philosophy is to 'show that the only nature that we will ever find worth preserving is one that has an inside and thus another side, [suggesting] a relation to something entirely "other" that allows us to find it not just valuable but venerable' (338).

In both versions, the nature's agency project tends to express an *ethos of respect* that decentres anthropocentric tendencies to perceive nature as a stockpile of resources for human exploitation. If respect centres attention on minimizing human agency, however, we should ask if this isn't problematic for people *already* decentred by the same power asymmetries facilitating the domination of nature. Does emphasizing such a complex, chaotic and transcendent world beyond human grasp and influence leave enough room to encourage grassroots political action—that is, movements empowered to challenge institutionalized forms of power resistant to the changes required by twenty-first-century issues?

All things considered, a 'dialogical partnership' model of relationality requires a third term to mediate social and socioecological relations. The philosophical demand is for a term sufficiently mutualistic to withstand the cultural pull towards polarization facilitated by the metaphorics of nature and agency. Politically speaking, this involves complementing the *ethos of respect* intended to keep human agency in check with an *ethos of empowerment* meant to encourage it.

THE PHILOSOPHICAL RESPONSE

Harking back to Husserl, Merleau-Ponty offers a philosophy of mutual relations mediated by the body as a 'third genre of being' (2012, 364), or what he later generalizes as 'the flesh'. The general thread running throughout his corpus, however, is a multidimensional articulation of what could be called *responsive relationality*. Here, then, we reconsider agency in light of the *response* as a viable third term.

If agency suggests an activity with interior 'mind-like' origins, the response is a 'dialogical' expression that emerges in the ambiguity and thickness of situated relationships. Importantly, for Merleau-Ponty, every situation *motivates* a response but doesn't *cause* it; and the response is *intended* but never simply *volunteered*. The body and past experience (personal and sociocultural) institutionalize background structures of meaning in ways that generally orient life towards a future to be realized. This background is what one 'brings' to a situation to make sense of it. (It's here that we locate Plumwood's interlocking dualisms and Merchant's root metaphors.) Yet for Merleau-Ponty, the situation itself is as important as one's background. The latter affords a field of possible responses to the situation in the process of perceiving, acting, thinking, speaking, etc.—but it's the contingencies of the situation itself that motivate the *appropriate* response. Contingencies are *generally anticipated*, depending on one's familiarity with such situations, but never specifically predicted. To some extent, there are always contingencies beyond our immediate grasp and influence that have to be 'questioned' (by, say, scanning for different motives in body language as we attune ourselves to a social situation). If, in the process of this embodied interrogation, certain phenomena stand out to meaningfully engage a possibility for response, the situation is 'taken up' and expressed accordingly. With each response, sense is made of non-sense, and the daily affairs of existence carry forward.

The dialogical response is situated and embodied and thereby ambigiously double-sided. It's 'passively' *im*-pressed by the 'external' contingencies motivating it and 'actively' *ex*-pressed by intentions 'internal' to one's background sensibilities. Hence, as *conditioned*, the response retains an 'internal' relation to the world and the past it stems from, without presumptions of organic unity. But as an intentional power *conditioning* the situation, it's also 'externally related' to a world and future still being created, with no question of 'mechanistic' separation.

Merleau-Ponty's paradigm for this ambigious relation involves a 'touching-touched' phenomenon whereby one of our hands touches the other. There's a distinction between 'actively' touching and being 'passively' touched, but there's also continuity because both hands belong to the same 'flesh'. We can also 'reverse' our attention to the touching hand itself being touched by the other. The ontological germ here resides in the 'crossing over' represented by such a switch. Here, phenomena traditionally counterposed become intertwined so that, at this juncture, the line between actively touching and being passively touched is ambiguous.

The same phenomenon applies to 'seeing-seen' relations. A painter absorbs himself in his work skilfully capturing the landscape from a distance, as it were. But he's equally absorbed *by* the landscape—feels himself in its midst, receptive to nuances of colour, texture and mood. One artist reports: 'In a forest, I have felt many times over that it was not I who looked at the forest. Some days I felt that the trees were looking at me, speaking to me. . . . I was there, listening' (Merleau-Ponty 1964, 167). The painter's movements thus seem to 'emanate from the things themselves', Merleau-Ponty explains, so that 'action and passion [are] so slightly discernible that it becomes impossible to distinguish between what sees and what is seen, what paints and what is painted' (ibid.).

How are we to understand this reversal between painter and landscape—this 'prehuman way of seeing things' that 'amplifies the metaphysical structure of our flesh?' (Merleau-Ponty 1964, 168). M. C. Dillon explains: 'The trees "see" in a manner comparable to that in which the mirror "sees" the painter: that is, the trees, like the mirror, let him become visible; they define a point of view on him which renders visible for him something that otherwise would remain invisible—his outside . . . his carnal presence' (1997, 161). But is Merleau-Ponty being metaphorical so that it's *as if* the trees 'see' the painter? And if so, why speak so literally about 'seeing' trees? Plumwood considers a similar issue in response to philosophers who argue that attributing mind-like qualities to natural phenomena 'as if' expressing intentionality produces better predictions. For Plumwood, 'as if' approaches that project consciousness onto nature are problematic because they refuse to 'take intentionality seriously in ethical terms' (2002, 184). Dillon makes a similar point. If Merleau-Ponty spoke of the trees 'as if' they were seeing the painter, this would tie him to the very philosophy of consciousness he intended to overcome (1997, 162).

Still, if humans respond to nature, and the response isn't anthropogenic but truly mutual, doesn't this imply that natural phenomena literally 'respond' as well? On the other hand, if we admit to being metaphorical, aren't we also forced to admit that only humans, and perhaps other 'higher' animals, truly respond—implying that everything else blindly *reacts* to things? Indeed, we glimpse why, as Jacques Derrida says, 'everything is in play in the distinction between reaction and response' (2008, 81). This is precisely the line of reasoning implicating Western philosophy in 'the discourse of domination itself' (89). What, after all, does and doesn't have the freedom and power to respond?

We glimpse here as well why, for Merleau-Ponty, 'a painting mixes up all our categories' (1964, 169). At stake are the dualisms we've been struggling with all along, and others like the metaphorical/literal divide. For instance, Merleau-Ponty states that 'our senses *literally* interrogate the things and that the things respond to them' (2012, 333. italics added). So when he goes on to say that 'nature must be our interlocutor in a sort of dialogue', he appears to mean it literally (334). And yet, as Jerry Gill argues, 'no other philosopher . . . has relied so heavily upon the metaphoric mode' (1991, 130). It isn't simply that Merleau-Ponty expresses *himself* metaphorically. Philosophical expression—indeed *all* expression—is inherently metaphorical. Phenomenologically speaking metaphor itself is an expressive gesture that, as Donald Landes explains, takes up and alters the meaning of something as an 'enduring response' to the weight of the situation that motivated it. Mediating our ambiguous internal-external relation to the world as we strive to get a grip on things, metaphor is 'always *between* and *around* the figurative and the literal' (Landes 2013, 37). Indeed, if you didn't pick up on the 'get a grip' remark as itself a metaphorical expression between the literal and figurative, perhaps this exemplifies how blurry this line is (ditto with 'pick up').

Returning to the 'seeing trees', then, we can now say that Merleau-Ponty was being 'metaphorical', but in a sense that intertwines the figurative and literal. As Ted Toadvine suggests, moreover, the 'striving' for expression, whether for the right metaphor or the right brush stroke, is also key to understanding something like nature's response. 'Nature offers to the painter a *motive*, that is, an embodied sense . . . [that] belongs as much to the landscape . . . as to the expressive movements by which he grants visibility on a canvas' (Toadvine 2009, 103–4). Being situated at the crossroads between intentionally absorbing himself in his work (seeing) and being absorbed by the landscape itself (seen), the painter paradoxically discloses nature 'on its own terms' but only to the extent that he takes up the landscape as his *own* (15). 'Nature, therefore, is precisely what discloses itself *through* our expressive acts, and as requiring such expression for its disclosure' (ibid.). The painting, we might say, is a visual 'metaphor' for the landscape, a response somewhere between literally representing the landscape itself and figuratively projecting the painter's own vision onto it.

Heidegger, too, captures a 'co-responding' relation with things, something John Llewelyn characterizes as speaking 'in something like a middle voice, poised between the putting of a question and being put into question,' and 'where creative power is inseparable from . . . responsive attention' (2004, 11, 18–19). The upshot for ecophenomenology here is, as Bryan Bannon says, that 'liberating beings from . . . [human] mastery is not a matter of disengaging from them, but rather that beings must be engaged with in such a way that we remain open to their mystery' (2014, 156). This includes being open to how places respond to our actions as we learn what socioecologically enhances them and what limits 'the ability of places to respond to further changes' (Bannon 2014, 164).

THE POLITICAL RESPONSE

The political counterpart of the philosophical demand for a third term concerns the need for nonpolarized expressions of respect and empowerment. If an ethos of respect encourages an attentive listening to nature and others, an ethos of empowerment invites one to find her voice and speak with intention.

Respect, as articulated in the nature's agency project, is an openness to being *decentred by difference*—being compelled beyond our egocentric, ethnocentric or anthropocentric assumptions in an effort to *hear* the voice of human and nonhuman otherness. Yet, if empowerment expresses confidence, one also needs to be sufficiently *centred* to 'speak up'. To be centred is to mature into one's identity with an integrity that remains open but isn't uncritical and easily swayed, or too reverent, safe or docile in the face of adversity, injustice and power.

Without a respect for the world beyond the one that centres us, the authenticity marking empowerment can't develop. Likewise, without an empowered, centred identity, the security and confidence needed to open oneself to difference with respect will be missing. In healthy dialogue, however, decentring encounters with difference motivate responses that, in their very expression, *recentre* us again. As Merleau-Ponty says: 'We have an *experience* of an I . . . that is indivisibly *unmade and remade by* the course of time' (2012, 228). This is simply another expression of the ontology of ambiguity outlined above. We're decentred by contingencies beyond our immediate grasp and influence, but recentred to the extent that we're prepared to effectively question things and take up the situation as our own in response.

Lacking respect and empowerment, relations are *reactive* rather than responsive. When reacting to something, whether to an insult or a baseball flying for our head, actions tend to be fixed, cause-and-effect relations where the activity/passivity and interior/exterior distinctions are clearly defined. Unlike the dialogical response, reactions are unidirectional movements to *defend* one's identity from external threats. If enough fear, anxiety or other insecurities compromise the integrity and confidence needed to recentre oneself, decentring encounters with difference can be quite threating. Held together by tenuous threads, we spontaneously react to protect ourselves from becoming undone.

To the extent that political discourses express conflicts along lines spelled out by the metaphorics of nature and agency, perhaps reactive debates that dogmatically harden lines in self-defence are to be expected. Here too, without a common ground of respect for difference and the empowered identity needed to recentre oneself, discourses degenerate into muscled reactions intent on protecting ideological centres. The options here, mirroring the two-term logic of dualism, is to keep a safe distance via 'hyper-separation', neutralize the threat by aggressively 'incorporating' the other or allow oneself to be submissively incorporated *by* the other. Dualistically structured discourses therefore compromise communication and collective decision making. In political contexts, they jeopardize solidarity within movements and partnership coalitions between them. In this sense, overcoming the philosophical problem of dualism addresses the political problem as well. John Dewey puts it well: 'Consequences within philosophy as such are of no great import. But philosophical dualism is but a formulated recognition of an impasse in life; an impotence in interaction, inability to make effective transition, limitation of power to regulate and thereby to understand' (1958, 241–42).

AN ETHICAL ONTOLOGY OF DIALOGICAL PARTNERSHIP

Are we not guilty of reproducing dualism by privileging responsive relations over reactive ones? Or are reactions sometimes more appropriate than responses? Although phenomenology's focus on description over prescription offers promising avenues for overcoming dualism, a viable response to the ecological crisis requires taking in ethical and political considerations as well. In this respect, ecophenomenology can learn something from ecofeminists. Discerning when to 'listen in' or 'speak up,' for example, requires sensitivity to social structures and power dynamics. It may not be ethically defensible, for instance, to expect the marginalized poor to quiet themselves to hear what nature is calling 'humanity' to do. Uncritically prioritizing the voice of privileged groups, moreover, could reinforce status quo assumptions that silence viable alternatives. And in cases where hegemonic interests prohibit mutual dialogue, people *should* be compelled to 'react' in the form of political struggle.

Here, then, I make the case for an *ethical* ontology of dialogical partnership—one premised on an ecophenomenology of responsive and reactive relations, but one that is also inscribed with ecofeminist analyses of power. Being institutionally situated in a world marked by adversity and asymmetries of power, it's important to first determine whether responses or reactions are called for in the context of social and socioecological relations, and then judge what *kind* is appropriate.

Let's take as our example experienced activists confronting efforts by the fossil fuel industry to cast doubt on climate change. Should they respond or react to this situation? The position I defend here is that responsive relations are privileged over reactive ones. Dualism is avoided, however, by distinguishing the specific from general contexts of a given situation. I therefore submit that reactions are sometimes justifiable in specific situations, but only if motivated in *response* to a broader situation. Recall that, for phenomenologists, appropriately responding to the *specific* contingencies of concrete situations requires being *generally* prepared to make sense of them against a background of meaningful possibilities. In this political context, therefore, I argue that our hypothetical activists should *react* to the fossil fuel industry's campaign to misinform the public *in response to the historical situation* marked by climate change.

Let's assume first, then, that the activists accurately situate climate change in its historical context by recognizing its roots in modernity's institutional infrastructure. Furthermore, they continuously dialogue among themselves and others about how best to respond to this issue. Importantly, this includes discussing their own social privileges to minimize unconscious tendencies to impose their vision on others unjustly. Ultimately, they've learned to see themselves as participants of a collective praxis of acting and envisioning that emerges in a thickly mediated milieu of social and socioecological dialogue.

In this historical context, it seems safe to say that our activists are truly responding to climate change. That is, against the general background outlined above, they're in a good position to judge whether the political situation at hand warrants a reaction or response. If, for instance, the fossil fuel industry thwarts all attempts at mutual dialogue to avoid drawing attention to itself, strategically reacting may be the best general response to climate change possible. As is required for decisive political action in this specific context, the lines differentiating activity from passivity, internal from external (e.g., 'us' from 'them'), etc., are relatively—but never absolutely—clear and unambiguous.

How did they get to this point of action? Recalling the discussion above, the activists could hear the call to act and respond appropriately only because they were open to listening for difference and prepared to speak to this issue in response. In the first instance, they respected the deeper implications of climate change, even while sensing that this receptivity risked profoundly decentring their identities by calling their world and future into question. But this courage, in turn, signals that they must have been sufficiently confident in their ability to recentre themselves in time, thus empowering their 'secret decision' to plunge into this abyss. That is, at some point, they must have been adequately prepared to sense the possibility of voicing a viable response.

Deeply implicating the basic sociocultural institutions we rely on to make sense of the world and our lives in it, climate change might be the most dramatic test case for an ethos of respect imaginable today. Can we really open ourselves to 'hearing nature's voice' here? Or could it be that questions about climate change are so difficult to answer—so terribly complex, multifaceted and heavy with existential implication—that they cannot even be *asked*? As Karl Marx says, questions only present themselves when people have the means of responding to them. Considering that the threat of climate change has been known for decades, we appear utterly impotent to take up this issue and respond appropriately. Today we confront a historical situation that demands a postcarbon future created by new voices and new centres of identity to express them. Hence, hearing the call of climate change and responding appropriately represents an equally dramatic test case for an ethos of empowerment as well.

If our hypothetical activists are appropriately motivated to respond, at some point they must have found themselves *situated* by this issue, and 'internally related' to it as such. Instead of sugar-coating it, ignoring it or engaging in other forms of denial in reaction to this existential threat, they learned to expose themselves to the gaze of an uncertain socioecological future. Somehow not overwhelmed, they also managed to intentionally *situate themselves* in relation to it. As expressive beings 'externally related' to this issue, then, they found room to carefully work through some very hard realities. The point of response was therefore reached once they found themselves being pushed forward by the unexpected contingencies of pragmatic engagement, while simultaneously pulling themselves forward in the process of collectively envisioning new horizons of possibility worth striving for. Not unlike the 'visual metaphor' expressed in the painter's dialogical involvement with the landscape, responding to the greatest collective action problem we've likely ever faced will certainly require something like a 'political metaphor.' Motivated to respond by the weight of our historical situation and expressed in the power-infused contexts of social and socioecological dialogue, our activists necessarily find themselves engaging a postcarbon future that intertwines the literal and the figurative.

CONCLUSION

Although environmental philosophers have long promoted a new 'relationship to nature', perhaps not enough has been done to address the nature of relationships. Ecofeminism and ecophenomenology are a powerful but underutilized exception to this.

I've argued that getting to a point of collective action with respect to problems like climate change largely boils down to being accurately motivated by them (respect) and creatively intentional about how they're taken up (empowerment). Yet, if we're to be sufficiently situated to hear the call while also situating ourselves in the process of voicing an effective response, the metaphorics of nature and agency will have to be exhumed from Western consciousness. Without reworking the 'everyday metaphysics' that secretly determine how we make sense of social and socioecological relationships, the monological demands for unambiguous consistency will continue to polarize our thinking and being in the world at the expense of dialogical mutualism. On these grounds, then, we need an alternative to the concept of agency, not an extension of it to nature. The response, as an embodied and ambiguous third term of mediation, is the alternative put forward.

Yet, being thrown into a world where problems are global, systemic and swiftly carried by the same institutional inertia situating our own lives, how we are to mutually cultivate responsive relationships to nature and each other? If an ethical ontology is to capture this dialogical synergy, it should encourage us to become skilled negotiators with the ambiguities of existence as we strive to make sense of a changing world that we're both internally and externally related to.

For Merleau-Ponty, the 'paradox of Being' that enables us to make sense of nonsense is that we are both 'a thing among things and otherwise what sees them and touches them' (1968, 137). Echoing this insight regarding the double-sided ambiguities of existence, ecofeminist Charlene Spretnak writes: 'How can we absorb the existential paradox of the universality of processes throughout the natural world . . . that give rise to each of *us* unless we become experienced in the kind of expression that can handle paradox: skilful metaphor, multivalent poetry and wise narrative of mythic depth that spark the imagination and reveal the rich unfolding of cosmological possibilities?' (1991, 22). In this light, I suggest, a Merleau-Pontian ecophenomenology in partnership with an ecofeminism along the lines traced by Plumwood and Merchant would amount to something like a politics of response.

NINE

Re-Rivering Environmental Imagination

Meander Movement and Merleau-Ponty

Irene J. Klaver

Climate change–driven floods and droughts have put water back on the map. The cultural realm has been inundated with water: a flood of water-related books, advertisements, brand names, real estate ventures, art projects, movies—including a James Bond film. Water has even streamed into the stock market: water stocks, as in infrastructure projects, desalination plants, bottled water companies, have soared. Water has become a solid part of our sociopolitical, economic world and our cultural imagination. Every self-respecting city in the industrial world wants to *re*-connect to its river. Where this is often a real estate and chamber of commerce driven process, it also bespeaks a shift in what I call environmental imagination: a socially, culturally and environmentally *re*-valuing of rivers. I thematize this movement of *re*-connecting, *re*-imagining, *re*-valuing and *re*-vitalizing rivers, as a *re*-rivering of our cultural imagination.

Where rivers had been anchors of civilization since ancient times, they became backgrounded in the industrial era (Klaver 2012, 15–19). Polluted, diverted and dammed, rivers 'disappeared,' were literally paved over or dried up because of impoundments, or relegated to the unsavoury side of town. From the backbone of town they became its backside: the unhealthy and unsafe zones, poor people's areas, harbours, dumping grounds, sites of water on fire. In the second half of the twentieth century a shift in mentality emerged with the rise of an environmental movement, culminating in growing environmental legislation, such as the 1972

Clean Water Act. These developments facilitated a *re*-newed foregrounding of rivers in city planning. Rivers meandered back into the cultural imagination, which by now had become an environmental imagination (Klaver 2013; 2014b).

In the following I explore this movement as a meandering movement, a movement of the again and the *re*-. I first trace the value shifts in the notion of meandering and *re*-connect meandering to its river, the Meander River in Anatolia, the Asian part of Turkey. I then show how the movement of *re*-rivering and meandering, in the very *re*-surgence of the prefix *re*-, forms a paradigmatic movement of phenomenology and especially elucidates Merleau-Ponty's phenomenology and ontology as an inchoative philosophy.

I then turn to an exemplary case of urban *re*-connecting to one's river: the city of Los Angeles's *re*-vitalizing of the Los Angeles River. I show how in the *re*-conceptualization of the Los Angeles River as well as in its realization, meandering abound, as does the prefix *re*-.

At the heart of the matter are *re*-consideration and *re*-imagination. How do we *re*-consider our situation, *re*-imagine our future, how do we facilitate a change in mentality and foster an environmental imagination? I argue that the prefatory syllable *re*- is a crucial catalyser: *re*-think, *re*-connect, *re*-build, *re*-configure. The prefix *re*- conveys the creative capacity of meandering and is at the heart of phenomenology and environmental imagination. It signals movement, ongoing sociopolitical and intellectual debates, the necessity and audacity of beginning again and again.

MEANDERING

In a modern era of efficiency meandering has become synonymous with aimless wandering, ambling along a winding path, rambling through a long-winded argument—the opposite of goal-oriented, focused and straightforward.[1] Meandering has acquired negative connotations. From early modernity onwards river management projects engineered away the meanders; they were taken out, rivers were, where possible, stratified, or re-placed by canals—the straight lines accommodating horse-drawn barges, making it easier to regulate commercial traffic on precise timetables. Linearity has been the privileged paradigm of progress and its leading model of efficiency; its concomitant mind-set has been goal-oriented or teleological. Meandering, convoluted and seemingly undirected, is seen as not just the opposite of efficiency, but as being in its way.

In the course of the second half of the twentieth century, nonlinear systems became widely accepted in the sciences—physics, mathematics and engineering. Complexity, chaos theory and non-deterministic as well as stochastic nonlinear modelling became the state of the art in many fields, including behaviour of large-scale natural or social systems in ecology, economy and politics. Analyses of both practices and systems highlight the importance of field-dependency, of a larger context. A growing awareness of these dynamics opened the door towards a widespread acceptance of complexity in the cultural imagination and a revaluation of meandering. Meandering allows for a re-thinking of progress through complexity: Symbolically and metaphorically, meandering conveys the nature of the nonlinear; it bespeaks the often-unpredictable movements of coming and going, the curving back and forth.

Re-valuing meandering has a train of effects on a variety of concepts and practices. Meandering facilitates a different way of thinking about efficiency, acknowledging that it might be more efficient in the long term to take more time and explore possibilities, just as a river meanders through a basin. More than control, exploration drives innovation. Meandering as a method, as a mental strategy, privileges exploration: a messy process, learning from mistakes, and following contingent relations. Many human practices develop in sinuous ways: learning through failing, honing a skill, building experience, facing unexpected challenges, starting anew. Meandering foregrounds the searching in the notion of research. Meandering invokes a model of engineering in terms of ingenuity, a bricolage and tinkering that acknowledges and interacts with various kinds of knowledge and expertise, that is capable of adjusting itself to local situations and demands. Meandering holds much in common with *Metis,* a term describing practical, even cunning intelligence in Greek culture; we will come back to this later, but in short: *Metis* stands for resourcefulness, practical effectiveness and experiential wisdom. Homer's Odysseus is *polymetis*—crafty, wily and cunning.

Meandering invokes, elucidates and hints at a different imagination, another mind-set, a new epistemological model and a cultural and political framework that diversifies what counts as expertise, knowledge, politics, progress and efficiency. Meandering does not allow for simple analyses or reductionist geopolitical frameworks; it demands ongoing political deliberation. It bespeaks the sociopolitical necessity of taking time to explore terrains, to elucidate attributes, relations, problems and solutions, as a gateway to new constructs of imagination, to a capacity to aspire.

The movement of meandering echoes an ongoing beginning and reveals how beginning works. Beginning does not take place in a vacuum, is not *a creatio ex nihilo,* but is always building on past experience, and, at the same time, can entail a break with this experience. Meandering is close to the epistemological and ontological workings of Deleuze and Guattari's notion of rhizome and Merleau-Ponty's understanding of movement. The emergence and fading of the Meander River in the cultural imagination can itself be seen as a meandering: an appearance and disappearance of the very river that left its indelible mark on human culture by giving its name to the process in which it disappeared again. The self-referential character intensifies the complexity of the process.

THE MEANDER RIVER

The River Meander (Anatolia, Turkey) once formed a mercantile and military conduit between Europe, North Africa and Asia. Herodotus mentions the Meander's winding ways and Strabo has given us the meaning of meandering as wandering. The earliest mentions of the Meander are found in Homer and Hesiod between approximately 750 and 650 BC. Hesiod (1914) gives a mythological account of how the river is named after one of the sons of the Titans Oceanos and Tethys: 'And Tethys bore to Ocean eddying rivers, Nilus, and Alpheus, and deep-swirling . . . Strymon, and Meander, and the fair stream of Ister' (337ff).

Homer mentions the Meander in his historical and geographical account in the *Iliad*. After the main characters from the Greek force are introduced, a catalogue of the Trojan troops follows, including those from the Meander valley, which he describes as rather uncultured, barbaric: 'And Nastes again led the Carians, uncouth of speech, who held Miletus and the mountain of Phthires, dense with its leafage, and the streams of Maeander, and the steep crests of Mycale' (Homer 1924, 870). Not long after these accounts of Hesiod and Homer, in which the Meander region is portrayed as rather backwards, the Greeks settled the Ionian Coast including the Meander Delta. The Meander valley became the vital trading route between the Mediterranean and Asia and emerged as a region of high cultural significance.

We might call this the first meandering of the Meander River in the cultural imagination. From a 'rural backwater' to the most precious gateway to the east: 'vast caravans of wood, wheat and spices, marble and ivory' followed its course (Seal 2012, 11–12). Trade and armies traversed the basin. The city at the high headwaters of the Meander River, Dinar (Celaenae in the fifth century BC), was of strategic importance: its pass connected East and West. Xerxes's Persians headed west in 481 BC to conquer the Greeks; 150 years later Alexander the Great headed east from Macedonia to conquer the Persians. These classic power shifts between the East and the West kept meandering along the river that gave the process the name.

Near the Meander's mouth on the Aegean Sea was the prosperous port of Miletus. In the sixth and fifth centuries BC it was a cultural centre, booming and bustling with celebrated musicians, poets, engineers, mapmakers and philosophers such as Thales, Anaximander and Anaximenes. Aristotle called Thales of Miletus the first Greek philosopher. Thales considered water to be the beginning, an originating and guiding principle or *archê*.

The Meander River had created a fertile valley. However, the very same agricultural development that made the region prosper and provided food for military and mercantile caravans enhanced erosion and silt formation in the basin. In another meandering of history, the once so powerful harbour city Miletus became a landlocked town. Over the centuries, the Miletus Bay silted up with alluvial deposits from the very river that nurtured its importance. The economy of the once-prominent harbour city collapsed. Nowadays, the ruins of the city lie some 10 kilometers from the Aegean Sea.

Rivers carry water and silt, two crucial elements in the complex dynamic movement of meandering. Meandering implies sedimentation and reactivation: silt settles, water scours.

MOVEMENT IN MERLEAU-PONTY

Sedimentation and reactivation are the fundamental problem of philosophy, according to Maurice Merleau-Ponty. This might seem to be a surprising assertion for a philosopher—however, not in the context of Merleau-Ponty's 'Working Notes', which were added to *The Visible and the Invisible,* the manuscript that was interrupted by his death. In a June 1, 1960, note he offers a concise and dense description of his phenomenology and ontology:

> In fact it is a question of grasping the *nexus*—neither 'historical' nor 'geographic' of history and transcendental geology, this very time that is space, this very space that is time, which I will have rediscovered by my analysis of the visible and the flesh, the simultaneous *Urstiftung* of time and space which makes there be a historical landscape and a quasi-geographical inscription of history. Fundamental problem: the sedimentation and the reactivation. (Merleau-Ponty 1968, 259)

Merleau-Ponty *re*-discovered the philosophy that time and space are co-originary; later there is a simultaneous fundamental institution/inauguration/initiation/origination (*Urstiftung*) of time and space. Any temporal event has a spatial sedimentation; any sedimentation is partaking in time, changing, moving. *The* fundamental problem is the dynamic between the two: How does stability form and how does renewal emerge from this stability?

In his earlier work the theme of renewal emerged in his emphasis on the beginning nature of philosophy. In the Preface to the *Phenomenology of Perception* Merleau-Ponty invokes Husserl, stating that the 'philosopher . . . is a perpetual beginner. . . . It means also that philosophy . . . is an ever-renewed experiment in making its own beginning' (Merleau-Ponty 1962, xiv). At the end of the preface he emphasizes the inchoative atmosphere of phenomenology, its unfinished nature, and summarizes: 'True philosophy consists in relearning to look at the world' (xx-xxi).

The theme of the re-, the again and again, emerges also when in the fourth chapter of *The Visible and the Invisible,* 'The Intertwining—The Chiasm' he sets out to develop his ontology. The chapter opens with stating,

> If it is true that as soon as philosophy declares itself to be reflection or coincidence it prejudges what it will find, then once again it must recommence everything . . . and install itself in a locus where they [I. K.: reflection and intuition] have not yet been distinguished, in experiences that have not yet been 'worked over,' that offer us all at once . . . both 'subject' and 'object,' both existence and essence and hence give philosophy resources to redefine them. (Merleau-Ponty 1968, 130)

Merleau-Ponty's emphasis on the *re-*, the re-discovering, re-learning, redefining, re-commencing, can be seen as manifestations of the fundamental problem of sedimentation and reactivation. Understanding these movements as meandering movements elucidates Merleau-Ponty's thought.

Merleau-Ponty never wrote a phenomenological analysis of rivers, never focused on any specific river. Unlike Heidegger, who wrote extensively about the River Rhine and the Ister, as in his 1942 lecture course at the University of Freiburg, *Hölderlin's Hymn 'The Ister'*—this is the same Ister we saw in Hesiod; it was the old Thracian and Greek name of a part of the Danube River. For Heidegger, Hölderlin's poetry bespeaks the poetic essence of the river. Humans might dwell at rivers, but rivers are independent, keep their own spirit and ultimately remain an enigma: What the river's flowing means stays also for the poet a mystery. Heidegger's rivers are metaphysical: the river is a poet, the poet a river.

Merleau-Ponty's rivers are metaphorical; they reveal a phenomenological and ontological move or force. He regularly invokes the flowing movement of streams and rivers as the source of metaphor and bearer of creativity into its future. It is a movement which proceeds rhythmically—ongoing rhythms, already under way in bodily perception, holding the past in the present as the promise and possibility of the future (Klaver 2005, 41). This fluid rhythmic movement conveys the way of perception, depth, space and time, a way that is, finally, the way of existence. He explains that as the sleeper awakens, 'the momentum of existence toward others, toward the future, toward the world can be restored as a river unfreezes' (Merleau-Ponty 1962, 165). When Merleau-Ponty develops his ontology, the rhythmic movement of water emerges again, this time as paradigmatic for the relation between the visible and vision: 'as though our vision were formed in the heart of the visible, or as though there were between it and us an intimacy as close as between the sea and the strand' (1968, 130–31). This close relation, the relation between sedimentation and reactivation, is the one of ebb and flow.

Organic, elemental, geological, in general: material metaphors populate Merleau-Ponty's philosophy, especially when he sets out to think about the intertwinement or chiasm between vision and the visible, subject and object. Underlying these metaphors is a sense of fluid movement, water, rivers, sedimentation and reactivation.

Merleau-Ponty uses water movement to elucidate his thinking. His water metaphors reveal the movement of his philosophy, of his notion of experience and his analysis of operative intentionality, which I employ elsewhere in developing a notion of environmental imagination (Klaver 2013; 2014b). In the preface of the *Phenomenology of Perception*, Merleau-Ponty invokes Husserl's concept of 'operative (*fungierende*) intentionality', that he does not further explore at that point, but which will pervade his last work, the *Visible and Invisible* (Merleau-Ponty 1962, xvii–xviii). He returns explicitly to it in the 'Working Notes' where he describes his own philosophy as the necessity to 'take up again and develop the *fungierende* [operative, I. K.] *or latent* intentionality which is the intentionality within being' (1968, 244). Intentionality is not simply located in the subject, neither in the object, but operative between the two. For example, seeing a picture of a glass of water makes me realize I am thirsty. It shifts the attention from the individual subject as agent to, what I would call, a situational agency, that is, the co-constitutive force of larger cultural and material patterns (of being), the embedded practices that co-determine how we experience things. Similarly, I see the environmental imagination as an *operative* imagination, understanding operative in Merleau-Ponty's sense of the word.

The specific movement of meandering, the movement of specificities, illuminates Merleau-Ponty's movements of thought. A meander is not just an invocation of water going here and there, but it invokes or refers to the interaction of the water with its riverbed, creating its channel, with the geomorphology, as the rocks and sands that are already there come in touch with the water and its sediments: Meandering is the interplay of rock, silt, water, lay of the land. With every interaction it starts anew. What is there, what is given, comes back in specificity—exactly how much sediment has been carried, how much water, how much land is impacted. All of these interactions do matter, play a role, do contribute to a particular state of a particular movement, defining specificity. Understanding this movement elucidates Merleau-Ponty's philosophy of perception and ontology.

Meandering clarifies streams of thought in Merleau-Ponty, his sense of operative as a movement of flow and his conceptualization of philosophy as inchoative, that is, a permanent beginning. Let us thicken our description of meandering.

MEANDER AND METIS

The very twisting and wandering character for which meandering became so well known bespeaks a way of thinking that has been long ignored, belittled, even considered counterproductive, precisely because it connotes complexity and multiplicity instead of linearity and unity. In its polymorph character, adjusting itself to the circumstances, meandering is structurally comparable to the ancient Greek notion of applied or real-world intelligence, *metis*.

In Greek mythology Meander and Metis (*Μῆτις*) were brother and sister. According to Hesiod, Tethys and Oceanos had three thousand sons, river-gods or *Potamoi*, and three thousand daughters, the *Oceanid*, each of them patroness for a specific spring, river or lake. Only the foremost were mentioned by name—among the sons, Meander, among the daughters, Metis.

Metis started off as an important deity, the first spouse of Zeus representing wisdom, skill, craft and cunning—a highly praised combination. However, Zeus, fearing her powers and her offspring, swallowed her, but she had already conceived Athena, who was born fully armed from his forehead. Metis faded from Greek mythology, eclipsed by her daughter, Athena, goddess of wisdom. Metis symbolized cunning intelligence in politics, practice-based knowledge in military art and medicine, the skills of the artisan crafts; all these forms of experiential wisdom, rooted profoundly in the intimacy of specificity, were called *mêtis*.

In *Cunning Intelligence in Greek Culture and Society*, Detienne and Vernant (1978) argue that *mêtis* is 'at the heart of the Greek mental world in the interplay of social and intellectual customs where its influence is sometimes all-pervasive' (3). However, despite its pervasiveness, *mêtis* is never explicitly thematized or analysed. While there are many treatises about logic, there are none about *mêtis*. The intellectual world of Greek philosophy, in contrast to the everyday mental world, was a dualistic world with a radical dichotomy between being and becoming, the intelligible and sensible, the unchanging one and changing multiple. In this framework of thought there was no place for *mêtis*, which 'is characterised precisely by the way it operates by continuously oscillating between two opposite poles' (5).

Odysseus is celebrated as the wily, cunning, the polymetis (*πολύμητις*), the one of many counsels (literal translation: *πολύ* 'many' *μητις 'mêtis'*). Yet the nature of his cunning, or of the skills of a craftsperson, the problem solving of a detective, the art of making a good joke, even the 'Eureka moment', went largely unexamined for centuries; indeed, many Renaissance thinkers scorned Odysseus precisely because he won by his 'tricks' rather than by applying rules and codes. *Mêtis*, often defined tersely as 'craft, skill, and experiential wisdom' came to stand for knowing *how*, for the palette of abilities outside the logical, goal-directed, static ways of knowing of *logos*.

Mêtis escapes simple definition—it 'always appears more or less below the surface, immersed as it were in practical operations' (Detienne and Vernant 1978, 3). Its way of knowing, its kind of intelligence and 'its field of application is the world of movement, of multiplicity and of ambiguity. It bears on fluid situations which are constantly changing and which at every moment combine contrary features and forces that are opposed to each other' (20).

'The essential features of *mêtis* . . . —pliability and polymorphism, duplicity and equivocality, inversion and reversal—imply certain qualities which are also attributed to the curve, to what is pliable and twisted, to what is oblique and ambiguous as opposed to what is straight, direct, rigid and equivocal' (Detienne and Vernant 1978, 46). Detienne and Vernant mention explicitly the term *skoliós* (twisted) in this context, as one of the adjectives indicating curving, frequently used to describe *mêtis*; the same term historian Strabo used two thousand years ago in his *Geography* to describe the Meander and meandering.

The Meander confounded early lawyers concerned with boundaries and scientists concerned with the mechanisms of meandering streams. Meander symbolized irregularity, complexity, ambiguity and instability. In the latter part of the twentieth century precisely these 'meandering' qualities brought out the value of multiple perspectives in arts and sciences; the weak ontology of becoming became as valuable as the traditionally more privileged strong ontology of being, the inductive, analogical and emergent as valuable as control and generalizability (O'Connor, Copeland and Kearns 2003, 99). The understanding of probability and complexity provided new forms of explanation and new ways to operate even within fields long founded on 'ideal' characteristics and laws. The meander came to be seen as an irregular waveform, at once subject to and generating random processes and forms.

Similarly, recent writers have begun to characterize emergent and analogical thinking. These characterizations of ingenuity bear deep resemblance to the *mêtis* of antiquity. Dreyfus and Dreyfus (1986) speak of *expertise* in terms of 'intuition [that] is the product of deep situational involvement and recognition of similarity' and note: 'how experience-based holistic recognition of similarity produces deep situational understanding' (29, 32). Similar concepts characterize modern ingenuity and engineering design: explicitly pragmatic; contingent; visual in character; satisficing; messy; holistic; whimsical; learning from failure (O'Connor, Copeland and Kearns 2003, 104).

Hapgood describes the first phase of engineering design as a 'metaphorical traversal through solution space', in which 'failure, imagination, and stuckness' are at play. The traversal and design process is 'idiographic and unpredictable' and often beset with 'painful trials or iterations'.

For Hapgood the engineer is a 'tinkerer who engages in activities within an artistic and subjective context' (Hapgood 1993, 96). O'Connor and Wyatt use the term 'thinkering' to blend Hapgood's tinkering together with Dreyfus's deep situational involvement into 'engineering discovery by doing' (O'Connor and Wyatt 2004, 12).

So we see *mêtis* acknowledged as a set of habits of expertise and the Meander holding its metaphorical power for today.

RE-MEANDERING

Re-meandering has become a popular practice in ecological restoration, even in places where there never were meanders. Rivers are resurfacing in the public imagination as cultural and ecological corridors, creating a cultural rejuvenation around urban renewal projects. In many rural areas river restoration is under way: The re-meandering of watercourses and restoring of floodplains are being carried out by many of the same engineering firms that straightened the waterways in the early or mid-twentieth century. New management regimes are seeking to work with, not against, rivers.

Meandering is dependent on the complex interaction of many material vectors. It is a symbol for how power operates in the everyday, lateral traversing, picking up material and depositing, re-activating in the process. Meandering stands for an ethics of adjustment, a politics of engagement, enabling and ongoing deliberation, a sense of experiment: tinkering, thinkering, emergent and transient. Meander brings the social, political, technological and natural together in an ongoing dynamic. The Law of the Meander is not the straight line but the sinuous back and forth, symbolized linguistically by the prefix *re-*, the notion of the again and again, the experience one gets in *mêtis*, the exploration through wandering, the essay in Montaigne's original sense of trial and attempt.

Meandering seems to be a slower process than the straight line of progress; yet this is only the case for the simply defined objective. Meandering proceeds by covering more ground, percolating into deeper depths, listening to the murmurs of more voices, being what it is when and where it is observed. Meandering makes room for what cannot easily be measured, what does not want to be measured, for the slow—slow food, slow movement, slow design—for the workings of the material realm not ruled by the structures of scheduled time.

The Meander River no longer functions as a Mediterranean thoroughfare, but the notion of meandering has re-emerged as valuable. Meandering is not a symbol for closure but one of ongoing change and exchange, of identities that shift over time. As Stuart Hall (1994) wrote: 'Cultural identities come from somewhere. But, like everything, which is historical, they undergo constant transformation. Far from being eternally fixed in some essentialised past, they are subject to the continuous "play" of history, culture and power' (396).

Meandering is a metaphor for the ongoing necessity of debate and negotiation in politics, a metaphor that bespeaks *mêtis*, the shape-shifting intelligence, which enables us to aspire to a complex future by beginning again and again. Perhaps no story better entwines re-rivering, meandering and beginning anew again and again, and environmental imagination than that of the Los Angeles River. A thick description of the recent history of the Los Angeles River sheds light on our discussions.

RE-IMAGINING AND RE-VITALIZING THE LOS ANGELES RIVER: RENEWAL AND RESTORATION

'There it is—take it!' So proclaimed and exhorted William Mulholland to forty thousand citizens of Los Angeles as water rushed into the San Fernando Valley on a November day in 1913.[2] This was not the water of the Los Angeles River. This water came from afar, from the Owens River Valley, carried across 220 miles of desert through the pipes of the brand new Los Angeles Aqueduct. Mulholland, who supervised the aqueduct's construction, gave the legendary and most concise dedication speech: 'There it is—take it!' (Mulholland 2000, 246).

The speech was the inauguration of an aqueduct, and of an era, a new mentality, a modern lifestyle. The new supply of water precipitated an era of explosive growth in Los Angeles, transforming it from a stagnating industrial town into a megalopolis, the City of Dreams. The trajectory of water and growth in Los Angeles is an exemplary case of the trajectory of modernity. Between 1850 and 1970, Los Angeles took a sheer utilitarian approach to water management, viewing water as a resource to be used as fuel for the urban 'growth machine' (Fulton 2001). The city embraced a policy of 'urban water imperialism', importing new water supplies from well beyond its city limits (Hundley 1992, 120). According to Reisner (1987), 'The Owens River created Los Angeles, letting a great city grow where common sense dictated that one should never be' (106). Water made Los Angeles—water mainly imported from rivers elsewhere. Waters of the Owens River, Feather River and Colorado River were re-directed in such great quantities that the Colorado River no longer reached the ocean and Owens Lake became a dustbowl. The Los Angeles River—the reason why the city was where it was—had been straightened into a flood-control channel, a 51-mile-long concrete scar (Price 2008, 547).

The Los Angeles River meanders throughout the story—a small unimposing river, outgrown, overgrown and initially straightened and ditched by the city to which it gave birth; then re-entering the cultural imagination as a character in its own right. The Los Angeles River presents us with a green-grey hybrid infrastructure that questions strict separations between human built/technology and nature, between various socioeconomic cultures and between different practices. This very hybridity gives promise of twenty-first-century urban rivers made to re-meander, re-creating a sense of water as commons and public space, which expresses a sense of culture as a capacity to aspire (Appadurai 2004). It is this re-rivering that puts rivers back on the map of our imagination.

In the 1980s Angelenos began to re-imagine Los Angeles's original water supply, the Los Angeles River. Many did not even know they had a river in their own backyard, even though they had, as most Americans, seen striking images of it—in the movies. In *Them!* (1954), James Arness encountered giant ants on its concrete banks; in *Grease* (1978), John Travolta drag raced across its paved bed; and in *Terminator 2*, future California governor Arnold Schwarzenegger escaped a killing machine from the future by fleeing down its concrete channel on a motorcycle (Gumprecht 1999, 244). People had no clue that this flood control channel was in fact a river. It was not experienced as a river.

In 1985, a group of Angelenos began to re-envision the Los Angeles River. Poet Lewis MacAdams became fascinated with the river. His work attracted other like-minded people, and together they founded the Friends of the Los Angeles River (FoLAR) 'to bring the River back to life' (Gottlieb and Azuma 2007, 27).

This resonated with an emerging cultural imagination infused with environmental consciousness (Klaver 2014a) and laid the groundwork for a grassroots effort to re-vitalize the Los Angeles River. Working through FoLAR, community activists began building bike paths, planting trees and lobbying for riverside parks, all with the intent of reclaiming the river, to make the Los Angeles River 'a place of community and ecological revitalization' (Gottlieb and Azuma 2007, 24).

By the turn of the twenty-first century, the movement to revitalize the Los Angeles River drew the interest of academics and city planners, who crafted plans for an urban revitalization program centred on the river. FoLAR partnered with the Urban and Environmental Policy Institute at Occidental College to produce a series of events called 'Re-Envisioning the Los Angeles River', which considered the possibilities of renewal from a multidisciplinary perspective (Gottlieb and Azuma 2007, 32–33). Voters approved bond measures for park and recreational development along the river, and the city established the Ad Hoc Committee on the Los Angeles River to coordinate the efforts of community activists, business leaders and city leaders to remake the river (City of Los Angeles 2011). Again, it was the inauguration of an era, of a new mentality, a re-envisioned lifestyle, this time the era of the river.

In 2007, the Committee issued the 'Los Angeles Revitalization Master Plan', which re-envisioned the river as the 'green spine' of the city. Renewing the river would renew the fragmented city: 'the revitalized River would foster community identity and civic pride, thereby bringing communities together' (City of Los Angeles 2007, 3–4). Its promotional brochure explains the plan's 'bold vision for transforming the River' and in large font with re- italicized: *RE*CLAIM, *RE*VITALIZATION, *RE*IMAGINE OUR RIVER, OUR FUTURE.

Once forgotten, the Los Angeles River re-emerged as a model of urban reclamation and sustainability. Revitalization promised 'flood control, cleaner water and cleaner air, and desperately needed neighborhood parks, wetlands, and wildlife habitat', not to mention increased 'local water supplies' (Price 2008, 551–52).

Even the U.S. federal government recognized the river's potential to re-create public space and re-forge communal ties. In 2010, the Environmental Protection Agency (EPA) officially designated the Los Angeles River a 'traditional navigable waterway', affording it additional protection under the 1972 Clean Water Act. The Los Angeles River and its community had removed the crust of concrete and invisibility and presented an exemplar of revitalization and renewal.

As urban planner and writer John Arroyo (2010) emphasizes in his thesis *Culture in Concrete: Art and the Re-imagination of the Los Angeles River as Civic Space*, 'artists have taken to the River as a creative venue. Their actions have re-defined the River and have allowed us (and impel us) to re-imagine the River as the civic space' (3). They have flourished in 'the un-designed, un-planned, and the spontaneous nature of the River space' independent of any formal urban planning or intervention (3).

River revitalization plans often come with gentrification and a complex re-drawing of the public and private: When the old abandoned, neglected, polluted, dangerous riverside becomes 'beautified,' poor neighbourhoods all too often are elbowed out to make place for a new upper-middle-class population. As Kibel (2007) states succinctly: With any of these projects one needs to 'consider the questions of who makes decisions about our urban rivers . . . and who ultimately benefits from or is burdened by these decisions' (15). The danger of commodification is a sanitized and controlled space—lacking the conceptual and social 'messiness' of abandoned and waste places, which function as meandering space. Foucault calls these places 'heterotopias': 'unique, non-traditional, and differentiated "other places" where the constraints of typical regulations and rules were suspended' and which therefore entice the imagination, spontaneous reactions and transformative powers (Arroyo 2010, 66).

It is too soon, yet, to say which ways the re-imagined Los Angeles River will flow. Price (2008) sums it up nicely: 'the revitalized river will be a product of continuous compromise and negotiation' (552). In fact, this is the definition of the politics of an engaged community that fosters the ongoing deliberative process of civic life, engaging in ongoing debate, meandering like a true river. In all its hybridity, the Los Angeles River crosses boundaries of race, class and human and physical geography, concrete and earth. As a re-meandering but still mainly concrete river it is the symbol of a new twenty-first-century paradigm of hybridity in water management and environmentalism. As such, the river provides a platform for the city to foster a sense of culture as a capacity to aspire (Appadurai 2004).

The ancient philosopher Heraclitus tells us one cannot step into the same river twice. The story-arc of the Los Angeles River enables us to reframe our relationships with our rivers, to revitalize the rivers and ourselves. In this renewed imagination re-rivered rivers meander back as experiential places.[3]

NOTES

1. The meandering and Meander River sections are based upon my previous writing on meandering. See Klaver 2014b.
2. This section is based upon previous writing with J. Aaron Frith on the history of water supply in Los Angeles. See Klaver and Frith 2014.
3. Acknowledgement: Thanks to Brian C. O'Connor for support, suggestions and close readings.

Part III

Practicing Phenomenology

TEN

Paradigms, Praxis and Environmental Phenomenology

Ingrid Leman Stefanovic, Zachary Shefman and Kristina Welch

> There is only one thing that a philosopher can be relied upon to do, and that is to contradict other philosophers.
> —William James, 'Remarks at the Peace Banquet'

That philosophy might offer more than a history of contradictions and theoretical debates described by James in 1904 is promising once one recalls Husserl's phenomenological dictum: 'To the things themselves!' (2001, 168). Reacting to a Western metaphysical tradition that for centuries prioritized universalizing speculation to the messiness of the lived world, phenomenologists from Husserl onwards arguably shifted the attention from merely abstract, scholarly debate to a different kind of engaged, philosophical commitment and understanding of being-in-the-world.

To be sure, the argument can be made that many phenomenologists and their postmodern cousins continued to principally argue concepts in a war of words. However, others paid heed to the call for a new kind of philosophical engagement in the *Lebenswelt*, illustrating in a concrete way how philosophy itself as a discipline deserves to be rethought in terms of everyday challenges of praxis and lived experience (Briggle, Frodeman and Barr 2015; Frodeman 2014).

This chapter explores the intersection of phenomenology and environment, specifically from the perspective of what it means for phenomenology to be 'applied' in an interdisciplinary setting. To provide some context, select examples illustrate how phenomenology has come to operate beyond the discipline of philosophy proper, informing non-philosophical enquiry from lived nursing practice to the design of built environments. A more focused, in-depth discussion follows, illustrating, in a practical sense, how phenomenological and hermeneutic methods are helping to uncover taken-for-granted paradigm shifts in our very understanding of the meaning of 'environment', specifically as it has informed four decades of university curriculum programming in Canada. We describe and critically reflect upon paradigm shifts that reveal changing values of interdisciplinarity and interpretations of 'environment', emerging from the early 1970s, when many environmental programmes were developed, to the present day. Ultimately, the aim of the chapter is to consider how phenomenology can move beyond traditional philosophical boundaries to impact our understanding of a broad range of environmental issues from policy development to education.

The first section addresses how phenomenology opens the door to a fundamental rethinking of the meaning of philosophical engagement in interdisciplinary settings. Over the years, interest has emerged in what some call 'phenomenology in practice', and certainly, there have been calls in the field of environmental philosophy for a different kind of phenomenological engagement in interpreting environmental priorities (Madjar and Walton 1999; Stefanovic 1987). How such moves reflect a distinct vision of philosophy than that caricatured by James, above—and how that vision calls for a unique notion of philosophy as 'lived' in the environmental field—is discussed.

The second part presents a case study, describing how phenomenological and hermeneutic methodologies have structured an investigation of shifting interpretations of interdisciplinarity, specifically within the field of what has for decades been described in Canada as 'environmental studies'. Interviews with key university leaders across the nation draw from a qualitative, phenomenological approach, in an effort to uncover strengths and weaknesses of interdisciplinary, environmental curriculum design. In addition, a hermeneutic investigation of post-secondary curriculum content from the 1970s to the present aims to identify paradigm shifts in our understanding of both 'interdisciplinarity' and 'environment'.

The final section then assesses the significance of phenomenology as it informs a different sort of environmental engagement for the discipline of philosophy itself.

PHENOMENOLOGY IN PRACTICE

In his magnum opus, *Phenomenology of Perception*, Merleau-Ponty poses the question, 'What is phenomenology? It may seem strange', he continues, 'that this question has still to be asked half a century after the first works of Husserl. The fact remains that it has by no means been answered' (1962, vii).

Martin Heidegger himself attempted to address the question, noting that 'within phenomenological inquiry, there are . . . differing definitions of its nature and its tasks. But even if these differences in defining the nature of phenomenology could be brought to a consensus', he mused, 'it would remain doubtful whether the concept of phenomenology thus attained, a sort of *average concept*, could direct us toward the concrete problems to be chosen' (1982, 2). Undoubtedly, Spiegelberg has it right when he notes that phenomenology, as 'on the way', is best described as a *movement* instead of a definitive doctrine (1960).

That being said, recently some have suggested that phenomenology, together with postmodernism, deconstructionism, poststructuralism and so on, remains within the realm of 'theoretical' approaches that 'litter' the field, serving as nothing more than 'windows with shutters, ways of seeing that close doors' (Denzin 1992, 22). In such cases, the conclusion is that 'standard theories are no longer working' and there is a need for a 'different approach'—perhaps one that experiments with new forms of 'visual and heard representations' (22, 26).

Our own sense is that drawing upon the notion of 'phenomenology' continues to be relevant, precisely because it was originally interpreted, as Heidegger notes above, as *other* than a universalizing 'average concept' or theory that closes doors and shutters thinking. Sensitive to the fact that experience is multilayered, complex, temporal and embodied, the classic term always did capture the need to understand our emplacement as ontological, nuanced and essentially mysterious. Moving beyond metaphysical hypostatization, recognizing that thought encompasses much more than a narrow rationality, acknowledging that experience is historical, changing and open to futural possibilities, phenomenology seems to continue to capture ways of being in the world that elude other speculative, reifying theoretical approaches to understanding.

Equally, phenomenology's focus on the *lived* world has meant that essential to its vision is to bridge the long-standing, metaphysical divide between theory and practice. To engage in phenomenological thinking is to illumine concrete, everyday taken-for-granted ways of being-in-the-world, rather than to fabricate theoretical constructs. That such structures of understanding are described less as *actualities* than as possibilities of encounter is captured well by Werner Marx, who reminds us that what makes studies phenomenological is:

> the way that they address the issues through an appeal to *experience*, not simply as a report about the author's own individual factual experiences or indeed necessarily about anyone's actual experience, but about the possibility of certain kinds of experiences which any reader should be able to recreate imaginatively on his or her own and thereby

> see that the possibility for such an experience is universal, even if the reality is not. (1992, 6)

That engagement in uncovering elements of possibility within *lived* experience suggests a new approach to philosophy proper. To some, it means a natural extension and 'application' of philosophy to the lived world but our reading of this engagement is quite different than such a top-down model might suggest. Instead, we see the intersection of phenomenology and lived experience to be essentially iterative and transformative of both philosophy and praxis. Martin Heidegger has it right when he notes that it is 'correct and proper' to conclude that we can't *do* anything with philosophy. At the same time, 'granted that *we* cannot do anything with philosophy, might not philosophy, if we concern ourselves with it, do something *with us?*' (Heidegger 1959, 10).

For a work to be phenomenological, a particular sensibility is required of a researcher—a willingness to absolve oneself of the comfort of theoretical certainties and immerse oneself instead in the messy business of how life interjects itself into our apparently neat categories of understanding. Neither a matter of 'applying' philosophy top-down to problems of everyday experience, nor simply an agglomeration of 'bottom-up' narratives that are mere inventories of plural perspectives, phenomenology seeks to navigate the space *between* mindful interpretation and a lived world that is given, rather than simply created. Such enquiry serves to illumine taken-for-granted truths, but those truths are seen as historically embedded interpretations of a world that is temporal and to that extent, never fully understood.

Perhaps because of this emphasis on a multilayered, lived world of human experience, the phenomenological method has informed other disciplines beyond philosophy. Questioning positivist approaches to their disciplines, critics have been moved by a recognition that phenomenology offers a possibility of exploring human experience in a unique way.

Psychologists, for instance, interested in moving beyond the fragmented, reductionist, behavioural modes of enquiry prevalent to their discipline, have turned to phenomenology to inform a 'constructive alternative in terms of praxis' (Giorgi 1970, cited in Kruger 1979, 113). Dukes (1984), Giorgi (1985), Polkinghorne (1989), Moustakas (1994) and others have explored ways in which the phenomenological sensibility translates to a different approach to psychology, one that is 'obsessed by the concrete' and whose primary aim is to 'observe, to comprehend, then to render explicit what was initially seen vaguely in the first comprehension' (Kruger 1979, 113). Acknowledging that we do not experience ourselves as 'pure mechanism, thing or mere organism', phenomenological psychologists aim to rethink statistical generalizations about human behaviour, describing instead unique, individual stories of human being as 'always already out there, interacting with others and the world' (Kruger 1979, 27).

A similar sensibility defines phenomenological work in the nursing profession. 'Interpretive phenomenology' is said to 'illuminate the world of the participants, articulating taken-for-granted means, practices, habits, skills and concerns' across a wide range of issues, from caring practices for hospitalized patients, to assisting families with critical care challenges, to advising patients who are recovering from illnesses. Through their daily 'bedside' experiences, care providers have come to realize that universalizing conclusions about complex, individual human needs of patients within plural environmental settings demand a different approach than that offered by analytical, positivist rules and standardized regulations. Providing assistance in such situations is seen to be better facilitated through a phenomenological approach to nursing that is guided less by a set of procedures and more through 'an ethic of understanding and responsiveness' (Benner 1994, xvii).

When it comes to issues of the 'natural' environment, phenomenology invites a different philosophical engagement from the standard analytic model of environmental and animal ethics. Drawing upon themes of embodiment, for example, thinkers such as Stephen Smith move beyond positivist, narrowly rationalistic arguments to explore the human-animal relation by attending to 'visible and invisible' actions, touches, senses and energetic transfers of interspecies connectivity' (2011, 9). By focusing specifically on the lived experience of training horses, Smith (a rider and horse trainer himself) explores the nature of a lived communion with animals through evolving training practices that now move beyond 'breaking' horses to more interactive, relational notions of 'horse gentling' and 'joining up' and ultimately, 'becoming other' [2011: 9, 15]).

Smith is one of a growing number of phenomenologists who are moving into the field of environment to rethink not only 'theory' but how thought itself is differently informed by way of an attentive, attuned engagement to the world as it shows itself, prior to theoretical constructs. Environmental Phenomenology has defined a new way of thinking beyond the traditional, analytic field of 'environmental ethics'. The International Association for Environmental Philosophy, an outgrowth from the Society for Phenomenology and Existential Philosophy (SPEP), was specifically founded in order to enable gatherings of scholars and students from the tradition of Continental thought to address the challenges emerging within the relation between human beings and their lived environments.[1] The affiliated journal, originally named *Call to Earth*, eventually became the refereed publication *Environmental Philosophy*, once again signalling that environmental issues include more than simply ethics but encompass a range of lived challenges emerging from fields such as aesthetics, epistemology, feminism, postmodernism, ontology, pragmatism and much more.[2]

Attending to built environments, the field of Environmental and Architectural Phenomenology[3] has emerged, with multiple publications about built environments, place, space and design (Norberg-Schulz 1979; Seamon and Mugerauer 2000; Malpas 1999; Casey 1993; Stefanovic 1987; Stefanovic and Scharper 2012). Significantly, in some instances, phenomenology has actually influenced architectural praxis. The impact of phenomenology here has been profound inasmuch as architects themselves have turned to phenomenology to inform their designs and to rethink their profession (Alexander, Ishikawa and Silverstein 1977; Baird 2015).

In such instances and more, phenomenology has inserted itself into interdisciplinary conversations that range from the architectural to health sciences. Arguably, in such cases, the impact of philosophy upon a range of practices, programmes and pragmatic decisions has been felt not only in terms of the *content* but the *methodologies* employed by these non-philosophical disciplines as well. In the following section, we consider examples of such new methodologies as they have inserted themselves into an interdisciplinary conversation between phenomenology, philosophy, lived experience and diverse examples of praxis.

PHENOMENOLOGY, HERMENEUTICS AND A CASE STUDY

Perhaps one of the most interesting examples of extending phenomenology into the field of architecture is undertaken by David Seamon, who builds upon Goethean phenomenological sensibilities to develop exercises to enable 'looking and seeing' of patterns and relationships within our cities (Seamon 2012, 242). Participants are invited to move beyond mere observation to 'plunge into the looking—they must literally "pay attention" so that they see *with intention* rather than just having a visual impression' (246). The hope is that through such a seeing, informed by phenomenology, 'we might assure a more natural city wherein at-homeness and serendipity, ordinariness and extraordinariness all have their place' (252).

Others within fields of social science have taken a similar inspiration from phenomenology to reframe standard data collection and interpretative practices. 'Qualitative', rather than quantitative methods have emerged as phenomenologically informed (Creswell 1998). Acknowledging that phenomenologists 'explore structures of consciousness', qualitative research methods aim to uncover 'the meaning of the lived experiences' without prejudging the content of those experiences (1998, 51). A major goal is to identify methods that avoid the imposition of rigid, theoretical presuppositions or statistically relevant expectations upon the investigations from the start.

The phenomenological approach to *interviews*, for instance, aims to incorporate an approach that is 'flexible and dynamic . . . nondirective, unstructured, nonstandardized and open-ended' (Taylor and Bogdan 1998, 88). Researchers within the social sciences have recognized that standardized questionnaires or detailed sets of interview questions already prejudge and theoretically frame a range of possible answers. The answers one collects are already shaped, inadvertently, by the rigid, predefined format of the questions asked.

Canadian planner Hok-Lin Leung beautifully illustrates this point with a story about how he was approached through a telephone survey for a chain restaurant (1992, 5–6). The interviewer was asking whether the 'all-you-can-eat' salad bar provided value for the money. 'You can choose one of the following answers', the interviewer offered. Explaining that his wife usually paid the bill, Leung asked the interviewer to remind him of the actual price of the salad bar. Citing slight differences among different cities, the interviewer replied that he was 'not supposed to give him the price'. Leung describes how:

> Undeterred, he asked if I would go to the same restaurant in future: 'Once a week, once a month. . . .' I answered, 'It depends. I don't eat out often but if I do, I would consider your restaurant. But I don't really know how often I will be eating out.' He said very apologetically that I must choose an answer from his list of answers or it would foul up his computer. (1992, 5–6)

Leung concludes that the experience was sobering: 'I must have designed many similar question-and-answer surveys in the past and had gotten numbers that were probably as useless as those I gave this hapless interviewer' (1992, 6). Quoting Terkel, Leung concludes that traditional, positivist survey techniques 'may be of value in determining favoured detergents, toothpastes and deodorants but not in the discovery of men and women. . . . It is foolish', he muses, 'to insist on unambiguous and logical answers. We must listen to all they say and then try to make sense of what we have heard' (7).

Seeking to enable such listening by avoiding the imposition of theoretical presuppositions, interviews are seen to be phenomenologically informed to the extent that they are conducted with minimal theoretical preconceptions, allowing for unexpected intrusions of the richness of lived experience to inform the research findings. In contrast to quantitative statistical surveys of the sort that Leung's interviewer was hoping to conduct, 'qualitative' phenomenological interview methods allow for in-depth understanding of both explicitly articulated, as well as implicitly understood, prethematic interpretive moments to infuse the interview process.

Typically, no more than 15–25 interviews are conducted with 'key informants'—although some argue that neither the number nor the type of informants necessarily needs to be specified in advance. Utilizing a 'flexible research design', the researcher 'starts out with a general idea of which people to interview and how to find them, but is willing to change course after the initial interviews' (Taylor and Bogdan 1998, 92). In the end, one is expected to use one's judgment to determine when adding additional interview subjects would yield no genuinely new insights (93).

Interview transcriptions are subsequently analysed as part of the interpretive research process to determine both spontaneously emerging commonalities among plural perspectives of interview subjects, as well as unique perspectives and insights that are allowed to emerge in individual cases, given the open-ended interview structure. Sophisticated tracking techniques are often utilized through computer programmes such as NVivo, allowing for the grouping of ideas that surface through the narratives.

Clearly, such unstructured interviews and interpretations that draw upon judgment calls of the researchers themselves call for a different sort of academic rigour than specified in quantificational methods of enquiry. On the one hand, the researcher 'as recorder and editor' of the interviews 'has a heavy hand in their production' (Taylor and Bogdan 1998, 135). At the same time, the aim is to not arbitrarily project one's own interpretation upon the enquiry but, on the contrary, to be open to the world, to listen carefully and gather information about a phenomenon that exceeds the parameters of rigidly predefined conceptual borders, in the hope of ensuring original and hopefully uniquely informative and nuanced findings.

At times, social scientists draw from Husserlean notions—experimenting somewhat controversially in my view, with the notion of epoché and bracketing. Phenomenological data analysis is said to proceed 'through the methodology of reduction, the analysis of specific statements and themes, and a search for all [*sic*] possible meanings. The researcher also sets aside all prejudgments, bracketing (see epoché) his or her experiences . . . and relying on intuition, imagination and universal structures to obtain a picture of the experiences' (Creswell 1998, 52). To be sure, phenomenological philosophers will find much that is problematic in such statements, from the notion that 'all' meanings are ever accessible to the notion that we can ever fully 'bracket' one's experiences. Nevertheless, it is noteworthy that references to Husserl's notion of 'bracketing' and intentionality are frequent within the qualitative research literature.

Altogether, the unstructured, 'qualitative' interview process employed within the social sciences is said to build upon a phenomenological sensibility that avoids positivist categorizations, essentializing theoretical constructs and reified research findings. Instead, by attending 'to the things themselves', the aim is to hear and interpret in a more fluid, open-ended fashion how unique life experiences are reported and understood within a context that is unconstrained by speculative constructs. Whether or not philosophers themselves see the merit in this exercise, the fact is that social scientists have built an extensive library of publications around such qualitative methods and, to my mind, those approaches would not have arisen without the contributions of phenomenology in the first place.

That phenomenology is core, then, to both the discipline of philosophy as well as other qualitative, social science research perspectives is clear. In these cases, phenomenology is seen to be less a distinctive, linear methodology than a sensibility toward interpreting lived experience—one that remains open to the unexpected, to plural realities and to temporality and change in understanding such experience.

A case in point emerges through an ongoing study, funded by the Social Sciences and Humanities Research Council of Canada (SSHRC). The title for the project is 'Interpreting Interdisciplinarity: The Case of Environmental Studies'. It was motivated by an interest in understanding and helping to resolve challenges relating to interdisciplinary teaching and research collaborations, particularly in the field of environment, which presents its own set of 'wicked' problems that defy disciplinary parameters (Brown, Harris and Russell 2010). It was also very much inspired by phenomenological and hermeneutic interests, which ensure that we seek to understand both explicit and implicit perceptions and world views that drive environmental decision making (Stefanovic 1987; Stefanovic and Scharper 2012).

In the case of this particular project, some standard information was collected in order to develop an inventory of Canadian university environmental units and to survey some general impressions about programme offerings across the country. In addition, however, several activities were undertaken that were explicitly informed by phenomenological and hermeneutic traditions. On the one hand, a set of interviews was undertaken, following the qualitative approach described above. On the other hand, we were also curious about paradigm shifts that may have occurred from the 1970s, when many university environmental programmes were developed, and so a hermeneutic exercise guided that set of activities.

Interviews were scheduled with more than 20 leaders of environment programmes in universities from coast to coast. A short set of questions was prepared in advance in order to help to kick-start each conversation. For instance, in addition to asking respondents to briefly describe their own offerings, the conversations typically included a question about perceptions of the meaning of 'environmental studies' and shifting interpretations over time. The interviews were conducted in a way that was spontaneous, unstructured and allowed for the respondent to lead the discussion in directions that they themselves felt to be significant. These narratives were then transcribed for purposes of interpretation and analysis. Even a survey that we sent out to about 100 university administrators and leaders (with a response rate slightly above 30 percent) was organized in such a way that questions allowed for extensive comments to be inserted, ensuring that a normally 'quantitative' survey invited spontaneous, 'qualitative' reflections from respondents.

In addition, a second strategy aimed to elicit evolving paradigms of interdisciplinarity and environmental programming. Drawing from hermeneutics, we were interested to see how language variations and changing interpretations of environmental concepts emerged over the decade from the 1960s to 1970s, when many university programmes were launched. As Stephen Smith points out, 'what we experience is inseparable from the linguistic framing of experiences such that what is seen, felt, tasted, touched, heard and smelled is already loaded with sense-making structure' (2011, 7). By collecting and comparing the language emerging from course and programme descriptions a decade apart, we were hoping to identify how different modes of expression reflected distinctive paradigm shifts in the interpretation of 'environmental studies' (as the broad, interdisciplinary field was originally called). Eight universities were selected from across Canada, ensuring that (1) their programming roots extended at least from the early 1970s to present day; (2) there was a certain prominence (quality and quantity) of environmental programming, compared to other schools; and (3) the study was geographically representative, drawn not only from the west to east coasts but also including francophone offerings from Quebec, given that both French and English are official languages of Canada.[4]

In this part of the project, we were inspired by similar work undertaken some years ago, where researchers aimed to identify paradigm shifts in water research through the framework of language shifts within a range of international environmental treaties (Bielak and Mount 2011). Beginning with the 1972 landmark 'Stockholm Declaration of the United Nations Conference on the Human Environment', eleven treaties were explored, ending with the Dushanbe Declaration on Water in 2010. Acknowledging that language shifts were often nondeliberate, the authors were able to identify not only a growing prominence of water issues within environmental treaties over the decades but also novel ways of understanding water security with respect to newly emerging issues such as gender equity and health. They were also able to critically reflect on meanings embedded in key terms: 'sanitation', for instance, is a word that may well distance us from the genuine, lived realities of the lack of plumbing and inadequate toilet facilities that plague the developing world.

Drawing from a similar intuition that linguistic changes reflect implicit paradigm shifts, we analysed transitions of language within university curricula over the decades, in an effort to understand how environmental concepts are differently understood and expressed today than when university-level programmes emerged in the late 1960s and early 1970s. Course as well as programme content was evaluated over all five decades using a variety of methods. Analysis was done on the complete course listing as well as subsets of the data, including examining only those courses that spanned at least four decades, exploring changes in wording across introductory courses and examining changes in required course requirements at the programme level. Finally, a holistic approach was taken to identify courses that reappeared over the lifetime of a programme, meaning that not only the course content but other factors—such as whether they were required or optional, whether they were prerequisites and so on—were taken into consideration when identifying salient courses as research foci.

All courses were reviewed and coded, using key subject themes for all decades and all disciplines. Word frequency over time revealed the top 100 words used within each decade, helping us to identify changing environmental priorities as revealed through shifts in curriculum content and design.

A number of findings emerged from the interviews and the hermeneutic deconstruction of programming shifts over the decades.

First, there were definite signs that the term 'environmental studies' has come to signify something different today for most—albeit not all—research participants. Significantly, when the term 'environmental studies' emerged in the 1970s, it was explicitly defined as an interdisciplinary approach that integrated the humanities, social sciences, sciences and the professions. For instance, page 13 of York University's *Faculty of Environment Handbook* for the academic calendar year 1970–1971 states that in establishing the Faculty of Environmental Studies, the 'focus of concern of the Faculty is upon man [*sic*] in relation to his physical, biological, cultural and social environments'.

In contrast, today more than in the 1970s, we hear university leaders actually distinguishing between 'environmental studies', on the one hand, and 'environmental science' on the other. One interviewee described environmental *studies* bluntly as 'a mixture of courses from a BA programme'. Others used the word 'policy' to characterize the field. A number of interviewees remarked that environmental *studies* is broader than environmental *science,* focusing more on the intersection between the 'physical' and 'social components of the environment' rather than merely the 'physical component'.

This emerging distinction is also reflected in the course content and arrangement within programmes. Where environmental science courses were previously (and briefly) included as part of the environmental studies programme, they eventually separated into their own degree or programme. Environmental science courses are now largely focused on earth systems, biology and other natural science disciplines, while environmental studies courses are focused on the human-nature relationship. In one case, a respondent felt that environmental *studies* has changed in the public eye. 'I believe they [the public] often see it as a non-specialized degree in which students are opting for an easy path—they do not see it as a real subject'. Environmental *science*, by contrast, was described more narrowly by reference to such terms as 'biophysical', 'technical' and 'quantitative'.

The fact that this broad notion of interdisciplinary 'environmental studies' from the 1970s has come to be understood as privileging the humanities and social sciences more than environmental *sciences* is perhaps not surprising, given that this large, 'holistic' and in some sense universalizing notion of environment that emerged decades ago has shifted in other ways as well.

Since the 1970s, both the notions of 'environment' and 'interdisciplinarity' have become far richer, more mature, complex, problems-based and pluralistic in nature. For instance, broad-based courses in 'Environmental Planning' were offered in the early days at York University in Toronto. By the late 1980s, nuances begin to develop through offerings in 'Urban and Regional Planning', 'Recreation Planning' and 'Theory and Planning of Rural Land Use'. By 2009–2010, there is an even larger variety of entries that include 'Landscape Ecology in Planning', 'Community Planning and Housing', 'Land Use Planning and Law', 'Politics and Planning' and 'GIS Applications in Planning and Resource Management'—indicating not only a greater *quantity* of offerings in the broad area of environmental planning but also a far *richer* and more nuanced diversity of courses, reflecting the need to explore these issues in greater *depth*.

In the words of one respondent, programme changes over the decades reflect that 'a change has occurred in how environmental change itself is understood'. While previously environments were perceived as 'largely stable, predictable and manageable', today there has been a paradigm shift that recognizes that 'complex systems are unpredictable and hence more challenging'. As a result, 'discourse has broadened' and become more 'complex'.

In addition, interdisciplinary courses have become more *applied and problems-based*, focusing less on abstract *theory* and more on *skills-based learning*. For example, course offerings in environmental impact assessments of GIS skillsets became commonplace across most programmes in later decades. This shift to more applied or career-focused courses was also reflected in the surveys and interviews: One survey respondent felt that 'the holistic goals of environmental studies have been increasingly supplemented by more *career-directed outcomes* based in environmental science'. In our interviews, respondents expressed their views that a problems-based approach was 'the best way to bring students from diverse backgrounds into dialogue with one another'.

To be sure, the reorientation of academic courses to incorporate more learning outcomes and objectives that align with career needs may not be unique to environmental studies/science, but rather, it may reflect a broader academic shift that emphasizes closer relationships between the academic community and industry and a desire to produce students who are 'job ready'.

Finally, 'interdisciplinarity' itself seems to be understood today in unique ways, when compared to earlier decades. Interdisciplinary systems thinking seems to be less about constructing universalizing, integrative syntheses and more about acknowledging and respecting plurality, complexity and *difference*. As one interviewee put it, becoming an interdisciplinarian involves 'challenging the discourse or the way you articulate ideas'. Our survey revealed that more than half of the respondents identify modest to considerable *shifts* in the meaning of interdisciplinarity over the decades. While it has come to mean 'blending social and natural sciences to answer bigger, more complex questions', one participant felt that 'the questions are no longer the ones being asked 50 years ago'. In fact, our hermeneutic work confirms that new issues, from a growing international focus, to questions of sustainable development, environmental impact assessment and environmental ethics, have emerged only recently in the interdisciplinary environmental arena of university course offerings.

The qualitative analysis of the full range of data collected on this project is still in progress. However, what is clearly emerging is the importance of both the phenomenological approach to interviews—which has allowed for a rich set of narratives to emerge—and the hermeneutic analysis of curriculum content shifts over five decades. Both have been essential tools that enabled the study of environmental programming at universities in a particularly meaningful way.

FINAL REFLECTIONS

Traditional philosophers may find fault with these sorts of descriptions of phenomenology as it has arisen within the social sciences and qualitative research methods. In fact, as is the case with many cross-disciplinary experiments, there are instances when social scientists exhibit an obvious misunderstanding or overly superficial conception of phenomenology, particularly in terms of its Husserlean roots.

Nonetheless, we remain convinced that phenomenology has much of value to offer beyond the confines of philosophy proper. Or perhaps better put—phenomenology signals the need to redefine the role of philosophy in practice. Bob Frodeman's recent book, *Sustainable Knowledge*, presents a compelling argument in favour of 'Field Philosophy'—a 'problem-oriented form of philosophical practice that treats knowledge production and knowledge use as dynamically integrated' (2014, 108). Such an approach to philosophy signifies the need for replacing 'investigator-initiated, discipline-based research with research that is problem-focused, interdisciplinary and socially engaged' (109).

Environmental problems today are more than merely theoretical. There is no living being on this planet that remains unaffected by global environmental changes. Those changes consist of more than explicitly articulated events that we can quantify and scientifically describe. They require a deep understanding of the human-environment relation as lived. In that sense, phenomenology has an even more central role to play in helping to elicit taken-for-granted judgment calls and assumptions that affect the policy advice that we seek, and the scientific questions that we ask in the first place.

It is time to take phenomenology beyond academic, disciplinary borders into a more meaningful interdisciplinary engagement with the lived world. Should philosophers feel that social scientists may not have got it quite right in their interpretations of phenomenology, then this is all the more reason for philosophers to enter that interdisciplinary conversation themselves. After all, as Plato himself said in Book VII of *The Republic*, the place for philosophical discourse is 'in both public life and philosophy', rather than in simply one or the other.

NOTES

1. For more information, please see http://environmentalphilosophy.org/ and http://www.spep.org/.
2. For more information, please see https://www.pdcnet.org/envirophil/Environmental-Philosophy.
3. The long-standing *Environmental and Architectural Phenomenology Newsletter*, edited by David Seamon, is available online at http://www.arch.ksu.edu/seamon/eap.html.
4. The eight universities were the University of British Columbia; Simon Fraser University, BC; University of Manitoba; University of Toronto; York University, Ontario; University of Waterloo, Ontario; Sherbrooke University, Quebec; and Dalhousie University, Nova Scotia.

ELEVEN

Re-appropriating the Ecosystem Services Concept for a Decolonization of 'Nature'

Barbara Muraca

The ecosystem services (ES) concept has become a powerful tool in an increasing number of policy and research contexts, ranging from ecosystem functioning research to biodiversity protection through ecosystem accounting and sustainability policies (Jax et al. 2013).

Since its appearance it gained both enthusiastic attention and a sharp critique from different environmentalist groups. Critics claim that it reduces 'nature' to a merely instrumental consideration and opens the floodgates to its ongoing commodification.

While I agree that the ES concept is being dangerously used to pave the way to the commodification of 'nature', I claim that it also bears a unique chance for reframing our very understanding of the relation to what we call nature in a new and refreshing way. The ES concept can help questioning and reframing the modern ontology of 'nature' and the concept of the 'subject' as opposite and separated from it. However, the paradigm shift the ES concept might help implement needs a radical process of re-signification and re-appropriation of its meaning against the mainstream interpretation supported by neoclassical economics and functional to the capitalistic mode of overexploitation. Thus conceived, the ES concept might become an ally to the struggles of indigenous people who are fighting for the preservation of their proyecto de vida, their collective vision for a self-determined and sustainable life in the community.

ECOSYSTEM SERVICES: A NEW PARADIGM FOR SUSTAINABILITY AND NATURE CONSERVATION?

The concept of ecosystem functions has been in use for a long time in ecology, mainly to address the different processes operating within an ecological system and their reciprocal relations albeit without any reference to the meaning and value of such functions for humans (Gómez-Baggethun et al. 2010). It wasn't until the early 1980s that the focus clearly shifted towards the societal value of ecosystem functions (Ehrlich and Ehrlich 1981) and their essential contribution to human well-being (Daily 1997). In Daily's book ecosystem services are defined as the 'conditions and processes through which natural ecosystems, and the species that make them up, sustain and fulfill human life' (3). From this point of view, the biological processes at work in different ecosystems are assessed in terms of their supporting and enhancing the life of human beings directly (production of goods) or indirectly (regeneration and cleansing functions), including aesthetic and cultural appreciation. A few years later the ES concept entered the international political agenda with the Millennium Ecosystem Assessment (MEA), called for in 2000 by the United Nations and resulting from the collaboration of more than 1,300 experts worldwide. In the MEA's definition of ES the attention shifts entirely from ecosystem functions to their benefits to people:

> Ecosystem services are the benefits people obtain from ecosystems. These include: provisioning services such food, water, timber, and fiber; regulating services that affect climate, floods, disease, wastes, and water quality; cultural services that provide recreational, aesthetic, and spiritual benefits; and supporting services such as soil formation, photosynthesis, and nutrient cycling. (MEA 2005, v)

Along the history of ES the concept of 'services' has played a hybrid role between ecology and economics. According to Kosoy and Corbera a conceptual shift from an ecological terminology (functions), which describes a functional relation among phenomena in terms of causality, to a genuinely economic one (services), which refers to a utility flow in terms of instrumentality, has taken place with a consequent process of ongoing commodification (2010, 1229). Whereas in the early debate the perspective is clearly dominated by environmentally concerned ecologists seeking a persuasive language to express the alarming loss of biodiversity and the deterioration of ecosystems, from the MEA the meaning of 'services' is gradually incorporated into the specific economic semantic of utility and benefits to humans (MEA 2003, 128).

THE CRITIQUE OF ES: FROM INSTRUMENTALIZATION TO COMMODIFICATION

The paradigm shift in the language of sustainability and nature conservation towards ES has on the one hand found a wide approval from rather pragmatic conservationists and environmentalists, who consider the economic valuation of ecosystem losses a strongly effective language in persuading policy makers and the business sector to invest in biodiversity and ecosystem protection (see, for example, WWF 2013). On the other hand, several concerns have been raised against the increasing use of the ES concept in connection with economic accounting and market-based mechanisms, such as Payments for Ecosystem Services (PES) (Spash 2008). These concerns are well founded. Assigning an exchange value to single ES implies transferring their use-value into an exchangeable unity to be traded on markets, thus turning them into commodities. Commodification requires abstraction, quantification and monetization: something can work as a commodity only if it is abstracted from specific characteristics relating to its space, time and quality and reduced to a universal equivalent measured by money. However, dividing ES into quantifiable and monetizable unities veils the background conditions for their (re)generation and isolates them from the complex web of intertwining functional relations in which they are embedded, thus neglecting the biophysical metabolism that supports economic and social systems and the long-term consequences of cumulative depletion. Moreover, according to Kosoy and Corbera, 'in the case of biodiversity, trade based on a single exchange-value imposes a language of valuation that can lead to a hegemonic worldview of nature' (2010, 1232) that is imposed onto different languages of valuation as they are expressed for example by local communities and indigenous people.

Critics of the ES concept not only point out the strategic nonsense of using monetization language for environmental purposes, which puts nature conservation under the pressure of having to demonstrate economically valuable contributions to human well-being and possible returns on investment, but also resist the overall framework of ES, which for them implies the shift towards a merely instrumental consideration of 'nature'.[1] They claim that the very language of ES reproduces an anthropocentric and utilitarian model, according to which 'nature' is to be protected only insofar as it yields some utility to humans, while excluding any consideration of its intrinsic value.

Interestingly enough, the MEA itself is aware of the limitation of a merely instrumental approach and tries to move beyond a narrow utilitarian paradigm by identifying 'another set of values' that are constitutive of the sociocultural identity of people and therefore cannot be framed in terms of solely utilitarian preferences. In an extensive chapter on ecosystem valuation, the MEA tries to include the sociocultural perspective on the value of 'nature' within the ES paradigm by referring directly to Kant:

> The counter-utilitarian idea that there is a difference between preferences and values and that considerations of individual rights tempers calculations of aggregate utility was most clearly and powerfully expressed by Kant, who wrote, 'Everything has either a price or a dignity. Whatever has a price can be replaced by something else as its equivalent; on the other hand, whatever is above all price, and therefore admits of no equivalent, has a dignity. But that which constitutes the condition under which alone something can be an end in itself does not have mere relative worth, i.e., a price, but an intrinsic worth, i.e., a dignity' (Kant 1959 [1785]: 53). (MEA 2003, 142–43)

However, the MEA fails to address how these two sets of values are related. Its framework tacitly reproduces the dichotomy between entities holding instrumental values, which are in principle monetizable and substitutable, and entities holding intrinsic value, which are bearers of rights in terms of dignity. Accordingly, 'nature' is conceived as either a mere standing reserve (Heidegger 1977) at our disposal that yields utility and serves humans' purposes or as something entirely separated, distinct and independent from human lives, and valuable for its own sake. In both cases the focus of what is 'valuable' relies on single entities that 'bear' value in the mode of an attribute rather than on processes and relations, out of which value emerges (Muraca 2011).

ES AS A PARADIGM SHIFT TOWARDS A DECOLONIATION OF THE MODERN ONTOLOGY OF 'NATURE' AND THE 'SUBJECT'

Modern Ontology of 'Nature' and Its Discontents

The concept of 'nature' that supports the concerns of environmentalists and fuels the idea of inherent moral values of natural entities is based on the very same modern, Western ontology of 'nature' that has enabled the overexploitation and instrumentalization of natural resources. Such an ontology is the outcome of the scientific revolution of the seventeenth century and relies on the bifurcation between an objective, passive, independent world, that can be analysed in terms of causal relations and measurable qualities, and a subjective world endowed with interiority, creativity and self-determination (Whitehead 1967). The necessary outcome of this bifurcation was on the one side a fixed and objectified nature, which can no longer 'speak' except through the scientific channels, and on the other a free, self-creative and detached subject (Muraca 2007).

The mode of non-instrumental relation to 'nature' resulting from the bifurcation reproduces what Gernot Böhme calls the 'bourgeois aesthetics of nature', that is, the aesthetics of the bourgeois intellectuals who encounter, in their leisure time, 'nature' as the naïve, unreflected, wild, primitive, non-civilized, innocent other, that can be observed, contemplated or enjoyed at a distance, from the point of view of those who are not engaged in a working relation to and with it. As Böhme rightly remarks, the externalization and instrumentalization of nature and the aesthetic relation of contemplation and admiration, all that in the ES approach is termed 'amenity and recreation value' and often captured by the idea of the value of nature in and for itself, are the two sides of the same coin (Böhme 1989).

After the bifurcation, the modern concept of 'nature' has served as the monolithic world of objective relations, the never-changing background for the multiplicity of cultural expressions: These could find their common ground vis-à-vis a unifying 'nature' that delivered a no-longer-questionable horizon of truth beyond all cultural differences (Descola 2011; Latour 2004). Instead of giving voice to different modes of relations among humans and nonhumans in their own right and specific language, the Western ontology imposes a colonial 'nature' as the framework for reading other ways of understanding 'human-nature relationships'. The point here is not to dismiss the modern ontology of 'nature', but to analyse it as one mode of framing the understanding of the relations with the world, as one specific ontology among others. As Descola writes:

> Rather than viewing the cosmologies of non-modern peoples as false beliefs and anthropocentric projections, geared more or less convincingly to chunks of positive knowledge, it is preferable to treat them, like all our actions in the world, as a way of patterning our relations with all kinds of entities in which we discern specific qualities, entities that require in return forms of behaviour and mediation that are adequate to the nature we ascribe to them. (Descola 2011, 18)

Descola calls Naturalism the cosmological dualistic paradigm that organizes humans in different collectives 'which exclude de jure not only the whole compound of non-humans, but also, not so long ago, exotic or marginal humans whose enigmatic mores, and the lack of spirituality or moral elevation of which these were the symptoms, led to their being assigned to the natural domain' (2011, 22).

By simply reproducing the bifurcation of nature, modern, Western environmentalism fails to grasp its unreflected collusion with colonial patterns of domination. Consider for example the wilderness conservation narratives of the Global North[2] where 'nature' is seen as 'the other' of society, a place for regeneration from the fatigue of urban or commercial life and necessary mirror for our very self-understanding as modern, 'civilized' subjects. Especially in the U.S. tradition, wilderness has been traditionally defined as 'an area where the earth and its community of life are untrammeled by man, where man himself is a visitor who does not remain,' (Wilderness Act of 1964, in Callicott and Nelson 1998, 120). Here 'man' is obviously the white colonizer whereas First Nation people are placed on the side of the wild. This is not a story of the past: Currently all over the world indigenous and poor people are being evicted from their territories by a pernicious coalition between Western environmentalists and the new bourgeois middle classes of their own countries in order to protect wilderness areas from which any use has to be excluded per definition (Guha 2000). According to post-development scholars, both traditions of environmentalism originated in the Global North, the resource management approach and the nature and wilderness conservation approach, are tightly connected and reveal the same basic logic. The bourgeois, romantic environmentalism that opposes the instrumental use of natural resources is equally oriented by consumption: It aims at creating and preserving special places, islands of intact nature, where no use is allowed, where 'wilderness' is restored and human intervention kept at bay in order to consume them in the form of documentary films, exclusive journeys, imagination, photography, etc. Against this pervasive narrative, another form of environmentalism, called environmentalism of livelihoods or environmentalism of the poor, is at work in many parts of the world, embodied by the environmental struggles carried out by indigenous people and peasants to preserve their own way of living in relation to their territory with its life-supporting processes (Guha and Martinez-Alier 1997).

Could possibly the concept of ES help reconsidering and questioning the modern ontology of 'nature' and thus become an ally in the attempt at decolonizing 'nature' and the modern 'subject' alike?

First and foremost, the ES concept refers to processes and their conditions of regeneration rather than to single entities. When considering ecosystems, what counts are the complexity of a dynamic whole and the reciprocal and manifold relations within it. Of course, the ES approach itself stems from the ontology of modern natural sciences and their analysis of 'nature' as a measurable object. However, if—by following Latour—we move our attention from the narrative that scientists are expected to reproduce in their publications and that is embedded in the general framework of Western epistemologies to what scientists actually do in their research and daily practices, we are faced with patterns that seem incompatible with that very narrative (Latour 2004; Sayes 2014). For example, when talking about ES from the point of view of ecology, 'ecologists think in terms of population dynamics, food webs, energy flows, interactive behaviors, biogeochemical cycles, spatial organization across landscapes, and co-evolutionary processes' (Norgaard 2010, 1219) rather than focusing on specific entities and their external or mechanic relations. It is quite difficult to generalize from the features and relational complexity of one specific ecosystem to other similar settings elsewhere and therefore to derive some general patterns about how specific types of ecosystem as they are classified in the MEA might deliver certain specific services to support and enhance human well-being. Recovering the concept of ES as it has been used primarily in ecology—by looking at it from a Latourian perspective—makes it difficult to fall back on a universal understanding of the services that 'nature' as a whole delivers to human well-being in general. This is also an argument against the aggregation of ecosystem values into one quantitative unit of measure such as acres or money.

Choosing to go back to the sense of ES that ecologists make use of in their actual practices does not mean to accept the scientific explanation of ES uncritically, without considering its provisional and contested political relevance. Rather, as Latour shows, such an ecology is highly political insofar as it gives voice to the multiplicity of elements, beings and processes with which researchers in their fatiguing work constantly interact on the field. According to Latour they struggle to include within the environmentalist discourse the annoying, contradictory details about specific relational networks, the disturbing way in which too many voices refuse being neatly organized in a coherent narrative of the function of 'nature' for humans and resist being dislocated, or better, delocated, that is, taken away from their particular geo-ecological embeddedness and relational contextualization. Rather than cementing an indisputable background of truth, ecology thus intended turns into a political deliberation that opens the scope of consideration of what is (Latour 2004). Appealing to scientists, Latour exhorts them to narrate what they do and experience in their daily interaction with other 'actants' (acting, active, sometimes reluctant processes)[3] instead of trying to force it into the epistemological framework of the modern ontology of 'nature' (Latour 2013). The work on ES thus conceived can help both to divest the idea of 'nature' from its almost mystical aura in terms of something separated from society and to focus on multiple reciprocal relations between human and nonhuman agents.

Decolonizing the Modern Subject: A Relational Axiology

Although the ES concept has been framed in terms of instrumental relations, it could also lead to a radical reconsideration of the different modes of relations between human and nonhuman processes in a more complex way and thus support a relational way of considering the value of 'nature' beyond the dichotomy between intrinsic and instrumental values. As a fundamentally relational concept, it can thus challenge the other pole of the bifurcation, the other of 'nature', the human subject encountering it.

The idea of a mere instrument relies on the assumption of a subject who exists independent from it and employs it as a means to his own goals. However, as phenomenology teaches us, our primary mode of interaction with the world is not in the form of a subject-object relation, in which we observe, analyse, choose and employ things as objects or instruments to our goals. Rather, in our daily practices we inhabit a world of shared meanings and expectations, and encounter things and beings—even tools—as other inhabitants of this world. If we consider for a minute the immediacy of our daily practices instead of the mediated epistemological framework, in which we have learnt to classify others as subjects, objects, instruments, things, humans and nonhumans, we must confess that we do not move as subjects in a world of objects, but interact with other actants, be they humans or nonhumans, material or symbolic (Latour 1996; 2004). In these interactions we constantly discover and negotiate our needs, desires and goals instead of standing before a set of means to be employed according to some predefined goals.

To say it with Heidegger, our most original mode of being is dwelling in a world that is not an object to us, but a system (or context) of interdependent and interlinked relations, in which we already find ourselves and that constitute us. In the mode of 'use' we do not, at first, instrumentalize things, not even tools: Rather, within this system of relations, we interact with them as bearers of some (shared) meaning. Whenever we 'use' a tool we become another and are internally and essentially changed in a way that is not completely under our control (Latour 2002). From the background of an ontologically fundamental system of relations single functional links, including instrumental ones, can then be posited on the ground of that very prethematic texture in which we are embedded (Heidegger 1973). It is only through mediation and interruption that we detach ourselves from the overall relational horizon and start gawking (begaffen) at things as something separate, isolable and analysable.

The dichotomic language of the MEA that encompasses only intrinsic or instrumental values reduces all relations within and to 'nature' to merely instrumental ones, thus reproducing the concept of relation as functional links and the modern ontology of an independent and detached subject. In order to unfold the potential radicality of the ES concept, the axiological framework has to be enriched by introducing—as I have articulated elsewhere—the more general category of 'relational values', of which instrumental values are but a subcategory—only useful when with good reasons the system of relations might be provisionally neglected. I propose an alternative axiological matrix to frame the complexity of relations with 'nature' that include fundamental, eudaimonistic and instrumental values (Muraca 2011, 2010; Jax et al. 2013). Fundamental values refer primarily to the ontological-fundamental system of relations that grounds the very possibility for the subject to arise and establish functional or even instrumental relations to the world it inhabits. Of course, they can also encompass fundamental relations in a rather material sense that includes the basic biophysical conditions for life—and for human life—on Earth as something that obviously is not standing 'in front of us' as a possible instrument to employ, but as a condition for our very existence and creative development.

Similarly, eudaimonistic values do not only refer to romantic, bourgeois leisure and aesthetic experiences—a walk in the woods, swimming in a pond or climbing beautiful mountains. Rather and more properly, they refer to all those relations and processes considered as necessary for living a 'good life' as a collectively negotiated meaning and a political project. From this point of view, eudaimonistic values are not confined to subjective preferences but extend to the self-understanding of a community, its social imaginary and the very basis for the common life.

The idea of fundamental dependency connected with the idea of human finitude and embeddedness, as it is stated in many non-Western traditions, reveals this vague intuition of fundamental and eudaimonistic relationality. For example, the connection to the 'Land' (especially for indigenous people) is not a 'functional' relation to single entities, which are valued for the benefits they provide (Muraca 2011). Rather, it represents the overall relational system that constitutes the common territory shared by all its inhabitants and encompasses their ecological, cultural and social interdependence, while supporting their collectively negotiated understanding of living well together (Buen Vivir).

CONCLUSION: RE-SIGNIFYING AND RE-APPROPRIATING THE CONCEPT OF ES FOR A DECOLONIZATION OF 'NATURE'

ES can display a path for rethinking relations between humans and nonhumans in a radical way and can help us moving beyond the dichotomy between intrinsic and merely instrumental values. By bringing to the fore the very idea of fundamental, constitutive and eudaimonistic relations, it can challenge any narrow consideration of 'mere' instrumentality as the only way of framing relations to the nonhumans while at the same time questioning the model of an independent, detached and self-creative subject. By focusing on the complex interconnection between ecosystem services and human well-being, the ES approach is essentially relational and might help, if not to overcome, at least to discontinue the obviousness in the Western tradition of environmentalism of the opposition between 'nature' and culture/society.

However, the ES concept can pave the way to a paradigm shift that helps decolonizing 'nature' and the modern subject only if it undergoes both a process of hermeneutical re-signification and of active and antagonistic re-appropriation.

Consider as an example of re-signification the very concept of 'services', which has been mainstreamed by neoclassical inspired economics in terms of sheer instrumentalization and is about to win the struggle for interpretative control against heterodox economic theories. According to ecological economics instead, services are flows that can be used without jeopardizing the qualitative regeneration of the so-called funds from which they stem. Funds are considered 'agents of production', active and transformative factors rather than piles of disposable matter and resources and comprise the traditional production factors such as Land, Labour and (man-made) Capital (Georgescu-Roegen 1971; Muraca 2010). Other than 'stocks' that decumulate over time and whose flows are consumed (i.e., used up),[4] services are used in a way that keeps them in good conditions for being used again by others (humans and nonhumans) and in the future. In this framework, use does not imply sheer instrumentalization as services are reduced neither to mere means nor to the benefits they deliver regardless of their own specific characteristics and of the complex relational system in which they are embedded. From the point of view of ecological economics, considering ecosystem functions as services implies entering a more complex relation that requires taking care of their conditions of regeneration. As far as monetization and marketization neglect those relations and specific qualitative characteristics, they have to be rejected. Depending on the complexity of the ecological systems and the sociocultural relations and shared meanings attached to ES, monetization might prove to be useful in some circumscribed cases, while being rather deleterious most of the time (Frame and O'Connor 2011).

Feminist economics as well has been struggling to re-signify the concept of services. So-called reproductive services, delivered both by what we call 'nature' and by other humans (mostly women and people from subaltern social groups), are those that support the reproduction of society materially, socially and institutionally: (re)productive services[5] comprise childcare, education, care for our and other bodies, care for relations, as well as the constant regeneration of soil, air and water accomplished by living organisms and processes. Despite their being taken for granted, neglected and excluded from economic valuation, they are the most fundamental basis for social life (Biesecker and Hofmeister 2010). Rather than being mere instruments to human goals, (re)productive services belong to the category of fundamental relations that constitutes the very subjects benefitting from them. However, neglecting their essential contribution is necessary both for maintaining in power the modern narrative of the independent and separate subject, who entertains merely instrumental relations with what 'serves' his goals (Ferber and Nelson 1993), and for the specific mode of exploitation that fuels the capitalistic generation of value: Not only 'nature' but also care activities are considered as a free gift devoid of creative potential, the passive basis that supports production in the strict sense of the creation of surplus-value—from which of course (re)productive activities are excluded per definition. Their capitalistic commodification (as for example the PES regime suggests) transforms them from services generated by funds, whose conditions and time of regeneration have to be respected in order to keep their transformative activity going, into flows stemming from stocks to be consumed as long as they deliver a significant margin for profit. While the capitalistic logic reduces services to stocks and benefits, thus moving from use to consumption and (over)exploitation, feminist ecological economics liberates the concept of 'services' from its mainstream-economic semantic of instrumentalization and re-signifies them in terms of what we can call a care relation.[6] Rethinking ecosystem services in terms of care relations moves beyond the modern ontology of the subject as it stresses not only our care responsibilities towards them, but also reframes our (human) condition as beings in need of care, dependent upon the incessant activity of living beings restoring the basic conditions for our flourishing. Care relations are essentially reciprocal: They both convey and need care.

Second, the ES concept has to be reappropriated to become a weapon for the decolonization of 'nature'. Reappropriating the concept of ES means to struggle for a plurality of valuation languages (Martinez-Alier 2009) and forms of knowledge beside the modern scientific paradigm to include indigenous, local knowledge and what is nowadays termed as 'citizen science' or AMS (Activism Mobilising Science; Conde 2014). Ecosystem services are the services of ecosystems—as trivial as this might sound, it requires attention to the specificity of fragile and complex systems of relations that cannot simply be abstracted from their geographical rootedness.

Following Latour we might envision a fruitful alliance between the actual practices of scientists in reconstructing industriously the different complex relations among processes, active forces and more or less collaborating entities and the mode of inhabiting a buzzing world of different beings—both humans and nonhumans—with which non-Western communities have managed to come to terms in different ways.

A more relational and pluralistic understanding of what we call 'nature' based on a radical rereading of the ES approach might thus foster what the anthropologist Arturo Escobar calls a 'decolonial nature', an understanding of the pluralities of nature-culture regimes that obey different representations and support the intricate structure of a territory with its living and nonliving inhabitants (2008). In fact, many ES are what they are due to interaction with human activity and would not continue existing without it: They are not only embedded in specific geo-bio-physical, but also cultural contexts. From the point of view of the environmentalism of livelihoods, what is at stake is not very much 'nature', but the creative processes and cycles that sustain life in general and secure its reproduction for the community inhabiting the land and interacting with it. Indigenous people, small farmers and women from the so-called Global South struggle for the preservation of their 'proyecto de vida' (Escobar 2008): their collective vision for a self-determined and sustainable life in common. The Land is the space that sustains the life project of the community and requires shared, social control of the modes of appropriation, use of and relation to natural processes as the basis for food security, self-subsistence and autonomy.

This becomes even more challenging if we try to relate specific ecosystem services, abstracted from their actual context, to human well-being in general as something that could be defined from the point of view of nowhere. The link between ES and human well-being is a politically contested field. Here as well we can learn from the idea of Buen Vivir developed by the indigenous people of the Andes, who have shown through their struggles that well-being requires a process of collective negotiation about meanings, basic institutions and political settings that include the understanding of the relation to 'nature'. While the idea of Buen Vivir cannot be simply imported to other societies because of its specific sociohistorical context, what we can learn from the struggles in Latin America is the re-appropriation of the political nature of questions regarding the good human life liberated from its (neo)liberal privatization in terms of individual lifestyles. Buen Vivir shows that the way in which we understand the relation between society and what is considered to be 'nature' most fundamentally determines how we understand and foster well-being. This framework challenges the Western idea of 'nature' as something external to society and shifts attention to a radically different understanding of the relation to the 'territory' and all its inhabitants (including nonhumans). Autonomy, self-determination and participatory decisions about basic institutions are the meta-condition for a framework that regulates the relation between ES and human well-being.

The focus has then to shift from a correlation between specific ecosystem features ('nature') and specific components of well-being ('culture') to a more comprehensive view of the political shape we collectively choose to give to society-nature relations and to the social and political conditions for well-being.

Rather than following the (romantic) myth of wilderness and the tradition of nature conservation for its own sake (and for the need for nature as our mirror), a different kind of environmentalism could assemble also in the Global North around the idea of cooperation, Gestaltung, interaction, co-creation, care, transformation within the politically negotiated conditions of regeneration and reproductivity of ES.

NOTES

1. I write here the term 'nature' in quotation marks to anticipate my considerations about how the very concept of nature is a product of Western ontology that cannot be found in the language used by indigenous people to signify their relations with other nonhuman inhabitants of their common territory.

2. This term is used to address Western industrialized countries and borrowed from post-development literature. Other than categories like 'West' it does not only refer to specific paradigms, but also stresses the asymmetric relations of power and domination between different world regions.

3. Latour uses the terms agents or actants with reference to their being efficacious and making a difference in a network of relations, while avoiding the philosophical quarrel about freedom or agency as ontological attributions. His ethnological analysis focuses on practices and does not claim metaphysical validity. Just the opposite: against the metaphysical assumption of a bifurcated 'nature' in modern naturalism, the focus on practices shifts attention from the big narrative of what nature and humans *are* to what they *do* and to the multiple, multifaceted and polysemic interactions, in which nonhumans are framed as collaborating, doing or not something, affecting, retreating etc. By shifting attention from the general epistemologies to what Latour calls agency (i.e., the sphere of actual interactions and reciprocal actions and affection), we might thus find more commonalities with non-Western modes of framing the relations among different inhabitants of a territory instead of dismissing them as primitive or non-scientific (Latour 2013).

4. According to mainstream economics, capital is defined as a stock that yields a (utility) flow. The concept of natural capital reproduces the idea of 'nature' being a stock that decumulates over time to yield benefits to humans. Different types of capital are in principle substitutable regardless of their specific characteristics and reproduction conditions.

5. Biesecker and Hofmeister write the (re) in parentheses to stress that (re)productive services are rather productive, creative and transformative, although they are intentionally excluded from production functions in mainstream economics.

6. Think of practices like permaculture, in which the fertility of the soil is used without being reduced to its mere utility flow (which might be replaced in the short run by fossil-based fertilizers). In a Latourian perspective we might frame the modes of interaction with the different actants operating in the regeneration of the soil and their complex relations in terms of interactions, interdependence and care.

TWELVE

Indigenous Experience, Environmental Justice and Settler Colonialism

Kyle Powys Whyte

Environmental justice (EJ) commonly refers to the problem that people of colour, indigenous peoples, women and people with disabilities, among others, are more likely than privileged white populations to live in toxic environments that are bad for human health and community cohesion. The idea underlying this conception of EJ is that justice concerns how the distribution of certain environmental nuisances, such as pollution, or environmentally related harms, such as asthma in children, burden populations who already face multiple forms of oppression, from structural racism to systemic poverty. Environmental nuisances and harms are treated as so many objects or states of affairs for which social institutions bear responsibility either to distribute equitably or to strive to lessen and, if possible, eliminate. Social institutions include environmental and natural resources agencies, legislative and executive branches of government, nongovernmental organizations, legal codes, and enforcement processes, among other examples, that are accountable to human populations to protect human health and well-being (Mohai, Pellow and Roberts 2009; Shockley 2012; Shrader-Frechette 2002).

In this conception of EJ, injustice is rooted in how social institutions are structured and operationalized in ways that favour powerful and privileged populations. These same populations, in turn, influence the structures and operations of social institutions in ways that protect their own interests without taking responsibility for the disproportionate burdens on other populations and communities. The conception of EJ just outlined covers many important dimensions of the nature of injustice, especially the impact of social institutions on the distribution of environmental quality across different populations.

Understanding EJ in this way maps on to many important facets of how indigenous peoples have been affected by colonialism. Indeed, indigenous peoples suffer environmental injustice due to the development of settler nations such as Canada and the United States through military invasion, capitalist exploitation of resources and the anti-indigenous policies of central and subnational governmental units (e.g., federal agencies, municipalities, provinces and the like). Environmental degradation—through commodity agriculture, deforestation, damming, mining, industrial air, water and soil pollution and nuclear energy and weapons development—produce multiple intersecting environmental hazards for which indigenous peoples are put particularly at risk relative to privileged settler populations (Agyeman et al. 2010; Coates 2004; Coombes 2013; Grinde and Johansen 1995; LaDuke 1999; Weaver 1996; Whyte 2011, 2015; Wilkinson 2005).

Consider the Aamjiwnaang First Nation of 850 members that is situated within 25 kilometres of sixty-two major industrial facilities, from oil refineries to manufacturing, in a region often called 'Chemical Valley' (by Sarnia, Ontario). The resulting air quality issues affect First Nation members disproportionately: Some 40 percent of Aamjiwnaang residents must use inhalers, and asthma affects approximately 22 percent of children and 17 percent of adults. Chemical contamination 'interfered with the community's cultural life, affecting hunting, fishing, medicine gathering, and ceremonial activities' (Hoover et al. 2012, 646; MacDonald and Rang 2007). Clearly, the nuisances and harms faced by Aamjiwnaang First Nation are disproportionately bad. Much of the responsibility falls on Canadian social institutions that make it possible for so much pollution to be focused on an indigenous people for the sake of the development of the settler Canadian industrial economy and the ways of life of settler populations.

Yet indigenous peoples' EJ movements and scholarly work focus on additional dimensions of injustice beyond the responsibility of social institutions for the distribution of nuisances, harms and goods. For many indigenous peoples, I will argue, injustice also occurs when the social institutions of one society systematically erase certain socioecological contexts, or horizons, that are vital for members of another society to experience themselves in the world as having responsibilities to other humans, nonhumans and the environment. Injustice, here, involves one society robbing another society of its capacities to experience the world as a place of collective life that its members feel responsible for maintaining into the future.

I seek to show how this understanding of environmental injustice is highlighted in theories and research from the domain of indigenous peoples and settler colonialism (Lefevre 2015). Settler colonial societies, such as the United States and Australia, are societies that seek to permanently settle the territories of indigenous peoples instead of *only* exploiting resources in that territory and sending the resources and profits back to a metropolitan homeland. In other words, settler colonial societies seek to inscribe their own homelands over indigenous homelands, thereby erasing the history, lived experiences, social reality and possibilities of a future of indigenous peoples. Settlers seek to inhabit the territories they newly inhabit as if they are—so to speak—the 'indigenous' populations. Settler colonialism can be interpreted as a form of *environmental injustice* that wrongfully interferes with and erases the socioecological contexts required for indigenous populations to experience the world as a place infused with responsibilities to humans, nonhumans and ecosystems.

I will begin to outline these points in this chapter through engaging with some articles in the literature on indigenous EJ in relation to settler colonial societies. I will then elaborate on the idea that for indigenous peoples environmental injustice is rooted in one society's interference with and erasure of another society's way of experiencing the world as infused with responsibilities. While here I cannot even come close to elaborating a convincing phenomenology or theory of justice, I aim to highlight some key dimensions of how some indigenous peoples connect EJ and ways of experiencing the world.

ENVIRONMENTAL INJUSTICE AND INDIGENOUS EXPERIENCES OF THE WORLD

Moral Terrains

Indigenous social movements and scholarly work on indigenous peoples and EJ often focus on dimensions of injustice that have to do with how one society interferes with and erases the socioecological contexts through which another society experiences the world. For now, a socioecological context refers to any arrangement of humans, nonhumans and ecosystems that involves particular cultural, economic and political practices. A series of papers by Figueroa, Waitt and McGee explore environmental injustice in the Australian settler state through the experiences of tourists who visit Uluṟu (the indigenous name of the monolithic rock) in Australia's Uluṟu-Kata Tjuta National Park. The original people—Aṉangu people—and the Australian government co-manage this place, which is significant to the Aṉangu according to their own system of Tjukurpa law:

> According to *Tjukurpa*, the Aṉangu collectively bond together in place through *Tjukuritja* (ancestral beings) that shaped the landscape from a void, creating mountains and living creatures on journeys, whilst also

> leaving some of their spirit being in each. Individual responsibilities towards spirits present in particular sacred sites clustered around the base of Uluṟu and the potential to release their power are only shared on the basis of the right to know by the appropriate indigenous person (Layton and Titchen 1995). *Tjukurpa* as law establishes an indigenous terrain of power that both names the rock-place as Uluṟu and identifies socially infused biophysical processes, allowing them to read Dreaming tracks of their peoples on the rock face. (Figueroa, Waitt and McGee 2007, 253)

Uluṟu through Tjukurpa law creates a socioecological context through which the Aṉangu experience the world as replete with meanings of moral responsibility such as being accountable to one's ancestors, having responsibilities to take care of the land for future generations and respecting the authority of different knowers in their community. Owing to this significance, the Aṉangu people do not climb the rock, and they wish that others also refrain from climbing.

Yet for settler Australians and many other visitors, Uluṟu is primarily a tourist attraction that they desire to climb as a matter of national pride or personal interest, which creates the following conflict:

> Yet, the most prominent car park locates the visitor at the foot of the climb. The path, created with chain linked rail-supports in 1958 by the park's first chief ranger, Bill Harney, remains open and highly trafficked. Indeed, any visitor passing the foot of the climb is confronted with a moral contradiction. In the foreground stands an unmistakable billboard, conveying in many languages the *Tjukurpa* law against climbing; in the background stretches the evidence of colonial habits sketched across the spine of Uluṟu in the form of a singular scar, a moral pathway for future climbers to endeavor. The park must also commit resources to rescue climbers in danger, and ward off would-be climbers during inclement weather conditions of extreme heat and/or high winds. (Figueroa and Waitt 2008, 334)

The tourists, for some reason, do not even notice the Aṉangu's prohibition against climbing as something worth respecting. This lack of respect both interferes with and erases *Tjukurpa* law and its significance to the Aṉangu.

In this case, it is possible to conceive of the major EJ issue as the desecration of a sacred site by a settler colonial population. The settler population is able to influence social institutions so that the Aṉangu have no rights to stop them. Yet as the authors show in their papers, the Aṉangu do not police or forcefully prohibit climbing because such forms of enforcement are not part of their Tjukurpa law. First, the Aṉangu see acts of respect as significant only if tourists do so genuinely. Second, the Aṉangu see policing as of a piece with problematic colonial strategies. Third, it would be even worse if the Australian government enforced Aṉangu law; that would replicate settler paternalism. Figueroa and Waitt see this problem as one of a 'relational vision of environmental justice' because it has to do with how different societies live within and negotiate different socioecological contexts (Figueroa and Waitt 2008, 346).

For Figueroa, Waitt and McGee, there is a socioecological context underlying how communities interact with humans, nonhumans and environments according to their own societal, spiritual, moral, economic and political systems and conceptions of how different entities may have diverse forms of agency. Here, ecologies are entire systems that orient how living, non-living and spiritual beings and aspects of the world are expected to coexist through relations of reciprocity, responsibilities and other ties and bonds (Dotson and Whyte 2013). They call these 'moral terrains'. Moral terrains are composed of 'the web of value layered over places through discourses that establish normative practices and socio-environmental belonging' (Figueroa and Waitt 2008, 328). A moral terrain consists of values, normative practices and states of belonging that are socially, historically and environmentally informed. 'Moral terrains are inscribed onto bodies through affective responses' (Figueroa and Waitt 2008, 328). That is, bodies express moral terrains through affective responses. These affective responses should be understood as 'living, performative emotions that enable people to come *to know and build or negate relationships* with other human body-selves as well as non-human entities' (2008, 328; see also Dotson and Whyte 2013 for an analysis of these concepts). For Figueroa, Waitt and McGee, the environmental injustice of tourists climbing the rock represents an interference with and erasure of the meaningfulness of Uluṟu as a moral terrain for the Aṉangu. As a consequence, Uluṟu's role as a socioecological context for Aṉangu affective and experiential lives is undermined.

Systems of Responsibilities

In 'Honouring Our Relations: An Anishinaabe Perspective on Environmental Justice', McGregor seeks to interpret EI in terms of her perspective as an Anishinaabe scholar-activist who has worked on numerous water quality initiatives on behalf of First Nations living in the Canadian sphere. McGregor has worked on a number of water-related issues such as the problems of drinking water in Canada. She discusses the Walkerton tragedy, in which seven people died from contamination of drinking water in Canada in 2000. In a report, it turned out that First Nations communities do not enjoy the same level of protection as non-Aboriginal communities in terms of the quality of drinking water. In 2005, the First Nation community of Kaschechewan had been under a boil water advisory for two years because of E. coli, the Ontario government declaring a state of emergency (McGregor 2009, 36).

She begins with the idea that 'relationships based on environmental justice are not limited to relations between people but consist of those among all beings of Creation' (27). People-centred EJ for McGregor includes issues of 'cultural dominance, of environmental destruction, and of inequity in terms of how certain groups of people are impacted differently by environmental destruction from others, sometimes by design' (27). She contrasts this notion of EJ with an Aboriginal one from her Anishinaabe 'world view'. EJ is

> about justice for all beings of Creation, not only because threats to their existence threaten ours but because from an Aboriginal perspective justice among beings of Creation is life-affirming. Aboriginal authors such as Anishinaabe environmental activist Winona Laduke refer to this as 'natural law.' While people certainly have a responsibility for justice, so do other beings (e.g., water and medicinal plants). . . . In the Anishinaabe world view, all beings of Creation have spirit, with duties and responsibilities to each other to ensure the continuation of Creation. Environmental justice in this context is much broader than 'impacts' on people. There are responsibilities beyond those of people that also must be fulfilled to ensure the processes of Creation will continue. (McGregor 2009, 27)

While 'natural law' is perhaps not the best term given the history of its usage in the field of philosophy, McGregor is really referring to notions of Anishinaabe law to designate *systems of mutual responsibilities* that include humans and non-human beings and entities. She cites LaDuke and also Anishinaabe scholar Borrows on how to understand the inadequacy of environmental laws.

LaDuke argues that 'natural law, which has existed for thousands of years, is a source of justice that has served Aboriginal people for thousands of years and that can continue to do so' (LaDuke 1993; McGregor 2009, 30). Borrows claims that 'Aboriginal peoples developed spiritual, political and social conventions to guide their relationships with each other and with the natural environment. These customs and conventions became the foundation of many complex systems of government and law'. Borrows goes on to say 'If the Anishinaabek do not honour and respect their promises, relations and environments, the eventual consequence is that these resources will disappear. When these resources are gone, no matter what they are, the people will no longer be able to sustain themselves because . . . while the resources have an existence without us, we have no existence without them' (Borrows 2002; McGregor 2009, 30). For McGregor and her interlocutors, then, experience within a particular socioecological context is what gives rise to people experiencing themselves as being part of a system of mutual, interconnecting responsibilities.

For McGregor, the drinking water threat represents a larger erasure of indigenous systems of responsibilities:

> It has been shown time and again in the history of Aboriginal-non Aboriginal relations in Canada that environmental destruction (of forests, lands, animals, waters) threatens the existence of First Nations peoples. Water quality, then, is not just an environmental concern; it is a matter of cultural survival. Among Native peoples, water is recognized as the lifeblood of the earth (a living and conscious being). In turn, water is therefore the lifeblood of the people in numerous ways (physically, mentally/intellectually, spiritually, and emotionally). Water is integrally tied to the cultural survival of the people. First Nations activists who have formed alliances to advocate and 'speak for the water' are at the same time resisting the genocide of their people. (McGregor 2009, 37)

Environmental injustice, then, concerns interference with and erasure of the systems of responsibilities—based in socioecological contexts—that are integral to the fabric of McGregor's society. In terms of systems of responsibilities, she discusses how 'All beings have responsibilities to fulfill, and recognizing this contributes to a holistic understanding of justice. Our interference with other beings' ability to fulfill their responsibilities is an example of a great environmental injustice, an injustice to Creation' (McGregor 2009, 40). Summing these points, she argues that

> As an example, water is a living, spiritual being with its own responsibilities to fulfill. . . . Similarly, it can be argued that because of their intimate relationship with the land, any injustice to Aboriginal people is an environmental injustice to the extent that it impairs the ability of Aboriginal people to fulfill their responsibilities to Creation. Conversely, any injustice to the environment that impedes the ability of creation to fulfill its duties to Aboriginal people is an injustice to Aboriginal people. (McGregor 2009, 40)

Similar to the idea of moral terrains, McGregor, along with her interlocutors Borrows and LaDuke, discusses environmental injustice as a form of interference with or erasure of a way people experience the world as infused with responsibilities that form the fabric of their lives as members of collectives. In Figueroa, Waitt and McGee and McGregor, LaDuke and Borrows, 'law'—whether Tjukurpa or Anishinaabe law—is often used as a concept to describe these complex cultural, societal and political systems of responsibilities that guide people's relationships with other humans, nonhumans and ecosystems.

Collective Continuance and Ecologies

Indigenous scholars have advanced similar EJ themes to those just discussed across a variety of literatures in environmental studies and native American and indigenous studies. I seek to focus here on my interpretation of the idea that the presence of certain forms of environmental change and variability, whether triggered by pollution, landscape degradation or climate change, alter how people experience their place in the world as saturated with systems of responsibilities connecting different persons, animals, plants and ecosystems; these systems of responsibilities are supposed to endure into the future and serve to support the capacities of societies to adapt to socioecological change. In this section, I will describe in additional detail my own interpretation of how many indigenous peoples discuss environmental injustice as tied to their capacity to experience the world as imbued with responsibilities.

To begin, consider some cases of how indigenous persons have interpreted environmental injustice as a form of interference with and erasure of the ways in which they experience the world as part of *collectives* that are constituted by systems of responsibilities. Haudensaunee peoples, including the Mohawk Council of Akwesasne, have borne the brunt of some of the most severe industrial pollution in North America (Arquette et al. 2002). Hoover, a Mohawk scholar of environmental justice working on issues with pollution and the decline of Mohawk fishing, quotes a community leader:

> As Henry Lickers describes, the language and culture around tying knots in nets as well as the social interactions that occurred around the process of creating these nets are lost when there is no longer a use for those nets: 'People forget, in their own culture, what you call the knot that you tie in a net. And so, a whole section of your language and culture is lost because no one is tying those nets anymore. The interrelation between men and women, when they tied nets, the relationship between adults or elders and young people, as they tied nets together, the stories . . . that whole social infrastructure that was around the fabrication of that net disappeared.' (Hoover 2013, 5)

For Lickers, the pollution of the waterways his nation depends on for fishing interferes with and erases the fabric of relationships that connect people, nonhumans (e.g., fish) and places (e.g., rivers, lakes) through mutual responsibilities to continue practices that are vital to the continued survival of the nation.

Climate justice has increasingly become an umbrella issue in which indigenous peoples discuss many environmental injustices. For, as scientists have pointed out, indigenous peoples contributed relatively little to the burning of fossil fuels and deforestation that drives global climate change today, but are suffering many of the consequences (Wildcat 2009). Willox, a scholar working on psychological responses to climate change in the Arctic among indigenous peoples there, interviewed a young female hunter about her perceptions of change in the Arctic as the habitat for certain animals she hunts changes:

> I think once it starts affecting one family it will start with the rest . . . once it starts affecting a couple of families really hard, it's not usually affecting those couple of families cause they choose it, it just happens to

> be those are the first ones to be affected, and then affect reaches out right? So I think if more and more people can't be going to the cabin and can't be hunting and can't be dependently going on the land that they just start to see a community shifting not knowing what they're supposed to be doing, not knowing what you're good at, not knowing what your self-worth is, not knowing what you should be doing with your time. (Willox et al. 2011, 20)

In Willox's work with indigenous peoples in the Arctic, perceptions of the disintegration of the fabric of community life have many ramifications. Willox discusses how 'For many elders, increased weather variability and unpredictability also have an emotional and personal impact. For much of their lives they have been able to advise the people around them confidently about when and where to travel, providing weather predictions. As their skills no longer work, some elders are now less confident and feel sadness that their advisory roles have changed' (Willox et al. 2011, 82). For those persons Willox talks to, the rapidly occurring climate change in the Arctic interferes with and erases their senses of responsibility to themselves ('self-worth'), to their collective societies, to the environment and to future generations.

In these cases, environmental injustice cuts at the fabric of systems of responsibilities that connect people to humans, nonhumans and ecosystems. Environmental injustice can be seen as an affront to peoples' capacities to experience themselves in the world as having responsibilities for the upkeep, or continuance, of their societies. According to my interpretation, responsibilities refer to the reciprocal (though not necessarily equal) attitudes and patterns of behaviour that are expected by and of various parties by virtue of the different roles that each may be understood to play in the various relationships that make up their collectives. For Lickers, cited in Hoover's work, people of different genders and ages had responsibilities to one another for the construction of nets, a key dimension of their culture and economic subsistence. In Willox's work, the elders have responsibilities to advise youth, which was an important system of knowledge exchange and mentorship. For McGregor, protecting water quality can depend on whether people see themselves as bound in relations of responsibility to water, with water as well having responsibilities to humans and nonhumans alike.

Systems of responsibilities are the actual schemes of roles and relationships that serve as the background against which particular responsibilities stand out as meaningful and binding. That is, one's particular responsibility for having a role in the process of making nets becomes important against the backdrop of the fact that net making contributes to the community or nation enduring into near and long-term futures. Systems of responsibilities also provide the feedback loops through which members of a collective can evaluate whether certain traditions should continue, change or be allowed to expire. For many indigenous peoples in North America today, what we refer to as our 'traditions' are responsibilities that were developed many hundreds of years ago in response to the socioecological issues of the time. So the relationships constitutive of systems of responsibilities also contribute importantly to a society's being able to recognize and facilitate new or improved responsibilities (Whyte 2013). My interpretation, then, involves the idea that *operating* systems of responsibilities both maintain vital traditional responsibilities and engender innovative responsibilities.

To bring this discussion back to EJ, environmental injustice can be seen as occurring when these systems of responsibilities are interfered with or erased by another society in ways that are too rapid for indigenous peoples to adapt to without sustaining preventable harms. That is, certain external forces may occur so quickly that there is not sufficient time or resources for a group to adapt without facing significant harms that they would not ordinarily have faced. In another case involving threats from oil to indigenous peoples in the Yukon Territory, Norma Kassi, speaking of her Gwich'in community, says: 'We cannot, however, simply change our diet. If we were to change suddenly and start eating store-bought foods more, then disease would increase and our rate of death would be higher, because it would be too rapid a change, too much of a shock to our systems' (Kassi 1996, 80).

I use the idea of *collective continuance* to describe my interpretation of the idea of systems of responsibilities adapting without sustaining preventable harms. Collective continuance refers to a group's capacity to adapt to external forces, from naturally occurring environmental change, such as the medieval warm period, to more obviously human-induced changes at several scales, such as the transatlantic fur trade or pollution of a fishery by an automobile manufacturing plant. Collective continuance is relevant to EJ because if one society interferes with or erases another society's capacity to adapt to external forces, then the former society—in promotion of its own interest and benefit—can impose preventable harm on the latter society's members. Harms can range from early death, to excessive pain and health problems, to limits on self-determination, to the erosion of cultural integrity, to many others. For many indigenous peoples, plants, animals and other nonhuman beings, entities and places are understood as among those capable of experiencing harm. In the literature I have been reviewing so far, I would argue that harms are understood as arising in a context where one society puts stress on another society's ability to have operating systems of responsibilities.

Environmental injustice, then, in the sense described in this chapter, occurs when one society succeeds in interfering with or erasing another society's collective continuance. The former society does so wrongfully when it seeks to protect its own collective continuance at the expense of the latter society's collective continuance. There are many things I can say about collective continuance that I will not be able to cover here; but I do want to highlight the institutional, ecological and experiential aspects. Collective continuance involves both the operationalization of systems of responsibilities (and their capacity to both rely on past traditions and transform appropriately into the future) *and* the simultaneous development of social institutions that can protect the operations of these systems. For example, institutions can include those geared to contest direct hardships imposed by colonial and other human interference, anticipate future challenges for which a group's political, social, cultural and economic systems must adjust, and facilitate decision-making processes that can guide decision makers through the most difficult circumstances, such as forced relocation or the extinction of a significant plant or animal. Examples of social institutions used for protecting systems of responsibilities range widely, from indigenous governments to laws, policies and treaties, to forms of activism, to indigenous knowledges such as traditional ecological knowledge, to indigenous economic enterprises (Whyte 2013).

Collective continuance is a coupled socioecological concept that can be thought of alongside the idea of moral terrains and Anishinaabe law discussed earlier. That is, the systems of responsibilities and social institutions can only operate when societies can adjust to certain ecological dynamics. Here, I will discuss further some ideas that are similar to the idea of socioecological contexts that I referenced earlier. Systems of responsibilities operate within what I will call an ecology, that is, an ecological system, of interacting humans, nonhuman beings (animals, plants, etc.) and entities (spiritual, inanimate, etc.), and landscapes (climate regions, boreal zones, etc.) that are conceptualized and operate purposefully to facilitate a society's aptitude to make feasible plans for the future—from the design of social institutions to the evaluation of traditional and emerging responsibilities for meeting new challenges. For example, systems of responsibilities that are tied to fishing must operate within ecosystems that can be taken care of in ways that ensure those fish populations are sufficiently abundant. Social institutions such as elders' groups or indigenous science agencies (such as the Akwesasne Environmental Task Force) serve the purpose of protecting the fishery (Tarbell and Arquette 2000). Or systems of responsibilities tied to interpretations of Uluṟu require that people do not disrespect the rock. Co-management arrangements, such as that between the Aṉangu and Australia, should protect the significance of the socioecological context. Similar to the earlier analysis of Figueroa, Waitt and McGee on moral terrains, an ecology provides a context for people to experience the world as full of responsibilities and future possibilities for which it is worth their effort to plan actively.

The idea of an ecology highlights an experiential aspect of collective continuance. A collective's capacity to adapt depends largely on whether members of that collective see themselves as having a future in which it is possible to lessen or eliminate another society's interference with or erasure of that future. Here, future can be understood in the sense of looking forward to future generations, but also in the sense of the continuance of enduring nonlinear cycles. This is why for the Anangu, for example, the settler society climbing Uluru is a major affront to them regarding their affective experiences, for the rock represents a basis for thinking cyclically about the future. Willox's interviewees expressed losses of self-worth and a sense of enduring into the future. For this reason, a key part of collective continuance involves people's experiencing the world within a socioecological context that provides a horizon that motivates them to participate in a system of responsibilities. Environmental injustice can be seen, then, as an interference with and erasure of a way of experiencing the world as imbued with responsibilities.

In some of his other work, Figueroa's concepts of 'environmental identity' and 'environmental heritage' relate to the idea that societies' relationships to their environments are important parts of how they envision a future for themselves that accounts for the ecological and experiential aspects of collective continuance. He writes that

> An environmental identity is the amalgamation of cultural identities, ways of life, and self-perceptions that are connected to a given group's physical environment. . . . Environmental identity is closely related to environmental heritage, where the meanings and symbols of the past frame values, practices, and places we wish to preserve for ourselves as *members of a community*. In other words, our environmental heritage is our environmental identity in relation to the community viewed over time. (Figueroa 2006)

In Figueroa, we see the ecological and experiential aspects of collective continuance understood as part of a person's identity and heritage as a member of a collective. Figueroa sees EJ as a community's designing and implementing social institutions that advocate for and protect environmental identity and heritage (Figueroa 2013).

There are many examples in the indigenous world of what I at least would interpret as offering similar concepts to the institutional, ecological and experiential aspects of collective continuance. The Sustainable Development Institute (SDI) of the College of Menominee Nation describes what I would call an ecology in terms of six dimensions that are always in tension with each other. In the 1990s, members of SDI felt they needed a model that can deal with the uncertainty and complexity of climate change impacts as integrated socioecological issues that connect diverse aspects of indigenous nations and communities and experiences with colonialism, from government to culture. For example, SDI was concerned with better understanding what the Menominee had learned through having to resist colonialism, especially its adaptation from a tribe primarily known for harvesting species such as wild rice to one known more today for its sustainable commercial forestry.

To express such complexity and uncertainty, the model defines sustainability using six interrelated dimensions: (1) land and sovereignty; (2) natural environment (including human beings); (3) institutions; (4) technology; (5) economics; and (6) human perception, activity and behaviour (Dockry et al.).

According to the SDI model, sustainability is the process of maintaining balance and reconciling tensions within and among these six dimensions of sustainability. Because all dimensions are of the same size, the SDI model implies an equal balance that expresses sustainability. Yet this model does not mean to imply that there is a functional equilibrium or a 'natural' balance, which can be challenged scientifically. Rather, change is an explicit feature of sustainability expressed by the model. Therefore, change in one dimension will influence other dimensions in an iterative

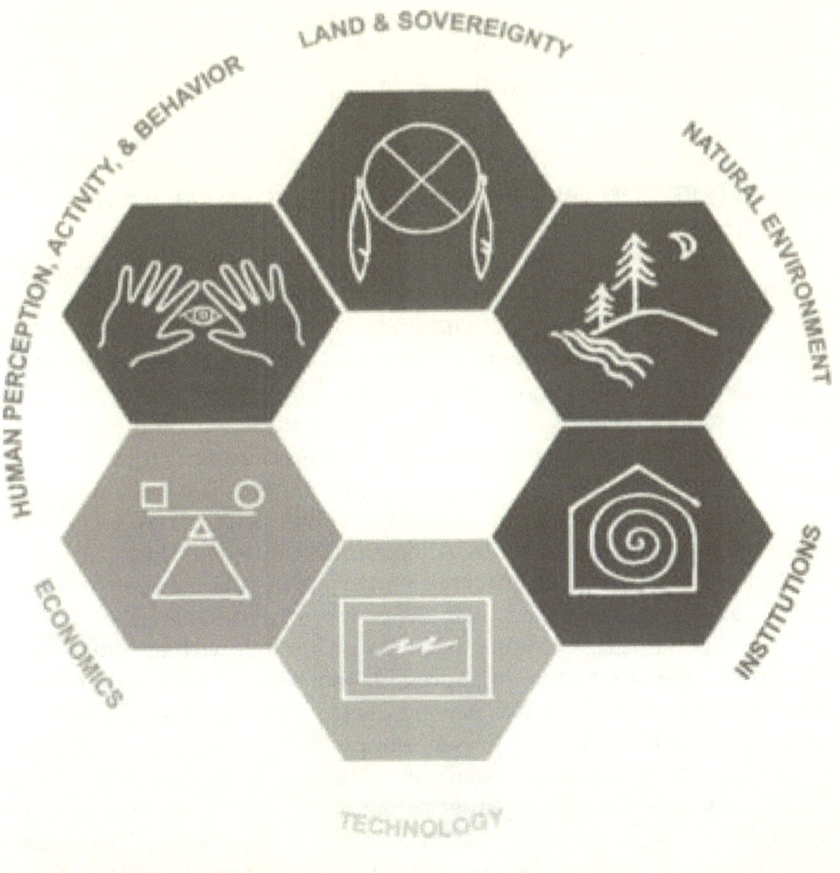

Figure 12.1.

cascade of changes. These changes are both externally driven and inherent to all of the six dimensions. Furthermore, as the model is brought back 'into balance', new tensions and states of disequilibrium will arise. Thus, sustainable development is a constant and iterative process (Whyte et al. 2014). I would say the same for collective continuance.

A key part of the Menominee model is the experiential dimension called 'human perception, activity and behaviour'. This dimension involves the idea that certain tensions can change how people perceive their relationship to the environment, have affective responses and feel motivated to behave. That is, for example, pollution affecting a fishery that lessens net making changes how people perceive their relationship to fish, to other people in their community and their levels of motivation for responding. So societies such as Akwesasne as described in the literature cited are finding ways to reinvigorate senses of stewardship, which may involve other trade-offs, if we follow the Menominee model, such as using new tools for fishing, investing in science education for tribal members or investing in language preservation programs. The Menominee model emphasizes how social institutions play a role in protecting what I am calling systems of responsibilities in the midst of tensions across the natural environment, economics, politics (land and sovereignty) and technology.

SETTLER COLONIALISM AND ENVIRONMENTAL JUSTICE

The institutional and ecological aspects of collective continuance are closely associated with the importance of how humans experience the world as a place imbued with enduring systems of responsibilities that they are motivated to continue. Widely cited concepts on the oppression of indigenous peoples connect very closely to this idea of environmental injustice and indigenous experience. I will conclude with some thoughts on how the concept of settler colonial oppression in particular connects experience and EJ in the ways I have discussed here. The literature cited in this chapter all arises from contexts of settler colonialism. I can only touch on a few aspects of settler colonialism in broad strokes given that ideas that have eventually become part of what this concept refers to today have been developed over many years by diverse scholars and activists (Lefevre 2015).[1] I hope to show in what follows that settler colonialism involves a structure of oppression that specifically targets indigenous people's ability to experience the world as imbued with responsibilities. Settler colonial theory presents important possibilities for further work elaborating phenomenologies and theories of justice related to EJ.

As a normative concept, settler colonialism refers to an arrangement of social institutions that support a structure of oppression. To understand the structure of oppression, we have to understand settler societies as those who relocate seeking to establish their own homeland in the homeland of another society. Here, I will use binary terminology, settler and indigenous, for expository clarity in this chapter, and will refer the reader to the literature cited for more complex accounts of settler colonialism. Settler societies are motivated to relocate and establish a new homeland for themselves as part of their own adaptation to external forces. So settler societies are, as any society, concerned with their collective continuance. Given the institutional, ecological and experiential aspects of collective continuance, settlers can only make a homeland by creating social institutions that physically carve their origin, religious and cultural narratives, social ways of life and political and economic systems (e.g., property) into the waters, soils, air and other environmental dimensions of the territory or landscape. That is, settler ecologies have to be inscribed into indigenous ecologies. Yet, for settlers, the indigenous ecologies testify to origin, religious and cultural narratives, societal ways of life and political and economic systems that owe nothing to those same aspects of the homelands settlers seek to inscribe. For settlers, indigenous socioecological contexts challenge settlers' claims to have honourable and credible religious missions, universal property rights and exclusive political and cultural sovereignty (Whyte 2015).

To expunge any markers or physical barriers challenging their legitimacy, settler societies engender social institutions designed to erase the socioecological contexts required for indigenous collective continuance. Settlement seeks to erase indigenous peoples' social institutions, ecologies and ways of experiencing the world for adaptation as a means of engraving settler social institutions, ecologies and ways of experiencing the world. In doing so, the goal of settlers is actually to eliminate themselves *as settlers*. Settlers seek to make another society's homeland *their* homeland, which literally involves making manifest the permanence and inevitability of the way they wish to experience the world in that place, from settler origin stories that seek to rationalize their arrival and development of land ownership to the political formation of their own political systems, from national governments to municipal and subnational governments, that serve to protect the homeland creation process. Indigenous peoples become entirely forgotten and invisible in the experience of settlers through settler educational systems that do not include curricula on indigenous peoples, laws and policies that ignore indigenous land claims, and cultural traditions that mock, romanticize and appropriate indigenous ways of life, among many other examples.

The physicality of the homeland inscription process of nations such as the United States and Canada cannot be deemphasized here, for the establishment of the social institutions of settler societies is interconnected with the very industrial activities that factor importantly into environmental degradation and anthropogenic climate change. Since settlers know little about the land that is new to them and hold cultural and economic values associated with their expectations for a certain way of life, the settler homeland establishes collective continuance through rather unsustainable means: deforestation, extraction, water and air pollution, commodity agriculture, urban sprawl, widespread automobile adoption and so on. These means are both built into the settler narrative of a homeland (e.g., narratives of entrepreneurs changing the world through industry) but are also sometimes hidden in plain sight (e.g., narratives of American wilderness mask histories of displacement of indigenous peoples). The industrial inscription of the settler homeland forms the background conditions needed for settler collective continuance in indigenous territories. Industrialization, then, is tied to settler ecological systems of relationships connecting humans, nonhuman beings and entities (often taken just to be inanimate resources) and collectives, such as forests cleared for agriculture, mountainsides stripped for mining and places of beauty separated from human non-recreational activities such as indigenous traditions (Whyte 2015).

Settler colonial theory, on my interpretation, offers a structure of oppression based on one society's interference with and erasure of another society. Given that the process described earlier is deeply ecological, settler colonial theory is always about environmental injustice. That is, settler societies seek to establish their own collective continuance at the expense of the collective continuance of another society. Settler societies impose preventable harms on indigenous peoples to facilitate the former's process of homeland inscription. The homeland inscription process replaces indigenous social institutions, ecologies and experiences with settler social institutions, ecologies and experiences. The cornerstone of this conception of EJ is its phenomenology of how people experience the world. A form of environmental change is unjust when one society robs another society of the institutional and ecological conditions required for members of another society to experience the world in ways that motivate their engagement in the operation of systems of responsibilities. While some accounts of environmental injustice focus on the distribution of environmental nuisances or burdens, this dimension concerns relationships between how societies negotiate different socioecological contexts.

Considering the themes raised in the literatures cited in this chapter, from Figueroa, Waitt and McGee to Hoover to the Menominee model of sustainability, understanding EJ in relation to systems of responsibilities, collective continuance and settler colonialism can serve to suggest very different solutions for remedying environmental injustices. Different from improving how social institutions distribute environmental quality, indigenous scholars, activists and allies often call for the resurgence of operating indigenous systems of responsibilities based on institutional, ecological and experiential aspects of collective continuance that can meet the challenges of negotiating settler colonialism today. I conclude by suggesting that those interested in the intersection of phenomenology and environmental justice can look to literatures on indigenous planning (Walker, Natcher and Jojola 2013), research methodology (Smith 1999) and resilience (Grossman and Parker 2012), among many others, for further insights.

NOTE

1. Settler colonialism has been linked explicitly to EJ (Hoogeveen 2015) and related concepts such as the relationship between health and historical trauma (Walters et al. 2011), education and visions of the future (Tuck and Gaztambide-Fernández 2013) and gender (Calhoun, Goeman and Tsethlikai 2007), among other connections. There seems to be an important trend of scholars exploring the connections between environmental justice, settler colonialism and important social and health issues. Scholars such as Emma Elliott, enrolled member of the Cowichan Tribes, are pursuing doctoral research at the University of Washington connecting suicide, environmental justice, and settler colonialism.

THIRTEEN

Music and the Presence of Nature

David E. Cooper

One of the great themes of the phenomenological tradition has been that of 'the human world'. This phrase conveys the idea that any world that can described and conceptualized is inseparable from the being of creatures like ourselves. We are 'beings-in-the-world' whose existence is not even notionally possible in abstraction from a world for us to be in. And the world, to be a world, must figure for or be present to creatures like us, for in the absence of the significance that things have in relation to our practices they could not stand out as things at all. Because it is a 'cradle of meanings', writes Merleau-Ponty, 'the world is inseparable' from our being (1962, 433). A world devoid of significance for us would be without structure and articulation, hence not a world at all.

The human world, in this sense, incorporates the natural world, for it is not simply the world of culture that is inseparable from human being. While people may have lost their 'immediate unity with nature', Hegel argued, it has been the 'labour of spirit' to recognize and 'restore the unity again' (1985, 42). For Merleau-Ponty, such is the interdependence of nature and humanity that 'it is impossible to say that nature ends here and that man or expression starts here' (2004, 319). Nature as described in the sciences—as so much material stuff—is, maintains Heidegger, a drastic abstraction from nature in a more fundamental sense: nature as it 'assails and enthralls us as landscape' and that, in virtue of 'our concern' has 'direction' (1980, 100).

The language of these philosophers often indicates that recognition of our unity with nature does not consist simply in assent to a philosophical proposition, but is something that it is important to have a feeling for. For Hegel, failure to appreciate this unity is a mark of 'the unhappy consciousness' that is not yet 'at peace with itself'. For Heidegger, to equate nature with so much stuff that stands over against us is not just philosophically confused, but serves to leave 'hidden' the nature that 'stirs and strives', 'enthralls us' and matters to us.

These philosophers agree, too, in identifying art as a way of cultivating a sense of our unity or convergence with the natural world. Hegel, for example, decries 'picturesque gardens' that represent nature as existing 'in itself', and insists that gardens should announce their artificiality and hence their relationship to human activity (1975, 248). For Merleau-Ponty, art—especially painting—is able both to 'draw upon' and communicate the 'fabric of brute meaning' that the world truly is, thereby redeeming it from the status of 'object-in-general' to which the sciences reduce the world (2004, 292f). In Heidegger's description of a Greek temple, this work not only 'opens up a world', giving to things their 'look' and 'to men their outlook on themselves', but conveys a sense of this world as 'set back again on earth' and thereby 'lets the earth be an earth'—not as 'a mass of matter deposited somewhere', but as the source and 'shelter' of human existence (1993, 168, 172).

Almost entirely missing, however, from these authors' treatment of the role of the arts in promoting a sense of unity with nature is a discussion of music. It is not mentioned by Heidegger among the Ancient Greek arts, like sculpture and tragedy, that play roles comparable to those of the temple, while for Merleau-Ponty music is 'too far beyond the world and the designatable' to draw upon and communicate meanings in the manner of painting (2004, 293). More generally in phenomenologically informed discussions of art and nature, music has been recessive. For example, the many attempts, inspired by Arnold Berleant's work (1992), to contrast people's aesthetic experiences of art works and natural environments respectively focus on the visual arts and rarely refer to musical experience. (An exception to this neglect is Gabriel Marcel—himself a pianist and composer—who records how both music and 'certain landscapes' have, for him, a similar 'power of incantation' [2005, 44].)

Neglect of the possible role of music in fostering a feeling for unity with nature is puzzling. After all, it is obvious, despite Merleau-Ponty's remark on music being 'beyond the world', that nature has been a main 'subject' of music. The themes of many 'tonal landscapes', moreover, are not nature 'in the raw', but human environments where people engage with nature. It is not wilderness, but the countryside that figures in Beethoven's *Pastoral* Symphony and Grieg's Norwegian Dances; it is a mountainscape in which men and women climb and hike that is sounded in Richard Strauss's *Alpine Symphony;* and it is the oceans that people sail and fish to which voice is given in countless sea shanties. There is no clear reason why music must lag behind painting in the capacity to 'draw upon' and depict 'the fabric of brute meaning' that is the human world.

In what follows, I want to redress this neglect of music and to suggest that phenomenology should draw closer to the concerns of the discipline of 'eco-musicology'. Since the phenomenological literature is so sparse, my remarks are, necessarily, something of a prolegomenon to the reflections that I hope to encourage. These are, in the main, reflections on how experience of—and with—music might cultivate a sense of our convergence or intimacy with the natural world. Musical experience embraces many ways of hearing and relating to sounds and some of these, I propose, are also ways in which nature may become especially present for us. Reflection on these ways, I also suggest, may enhance our understanding of what it is to have a sense or feeling of unity with nature, and of a 'codependence', as one might call it, of culture and nature. Talk of 'oneness with nature' is pervasive in some sectors of environmental philosophy, but it is difficult to construe this talk so that it is neither bland nor starkly false. Arguably, the slogan names an aspiration to live and feel in ways that are consonant with the recognition that the world is a human world. If so, then one strategy for understanding 'oneness' is by reflecting on experiences—including musical ones—that further this aspiration.

MUSIC AND EXPERIENCE

Reflection on musical experience requires an understanding of the richness of such experience. Asked if they like music, or if it matters to them, people usually take the question to be about listening to music, in venues like concert halls or on devices such as CD players. Music tends to be thought of in modern culture as, first and foremost, pieces of music that are composed in order to be listened to 'for their own sake', for the aesthetic satisfaction they provide. But as many anthropologists point out, this is a very 'late' understanding of music corresponding to a similarly late development in Western musical culture. Music, so understood, should be regarded as growing out of earlier musical practices. For Christopher Small, in an influential book, music as an object of appreciation is a sophisticated offshoot of an amorphous activity he dubs 'musicking', which embraces performance, listening, 'providing material for performance', dancing and much else (1998, 9). In early societies and many surviving 'traditional' cultures, the music typically didn't and doesn't consist of 'works of music', is not 'composed' and, crucially, is not performed 'for its own sake', but is integrated into wider purposive practices—rituals, say, or preparations for warfare or hunting. Far from being 'an art', to be enjoyed at leisure, music was, as John Dewey put it, an 'integral part of the ethos and institutions of the community'—one in which 'the idea of "art for art's sake" would not have been even understood' (1980, 8). When Confucian texts, for example, discuss music it is in the context of the role of music in the rites that, for the authors, were essential for the maintenance of social morality.

That 'musicking' had pragmatic advantages is presumably relevant, if not to the origins of music, then at any rate to explaining how it became an entrenched feature of virtually all early societies known to anthropologists and archaeologists. The advantages conferred by music on which scholars have speculated range from instilling a sense of community to serving as a form of emotional expression that is difficult to dissemble, and from enabling communication at a distance or with animals to reinforcing certain essential activities by making them more enjoyable. (See Mithen 2006 for these and other speculations.)

Anthropological and historical accounts of the entrenchment, development and functions of music are useful for exposing how limited and parochial is our dominant contemporary understanding of music. Useful, but not perhaps essential, since it should be apparent to any alert observer that, even in twenty-first-century Western societies, people's engagement with music is much richer and more diverse than that conception would suggest.

The relevant point can be put in terms of the notion of experience. People whose understanding of this notion as shaped by the Empiricists or by Kant are liable to regard experience as the reception of sensory data—or as this plus the mind's organization of the data under concepts. But this is a notion that fails to capture what is intended when speaking, for example, of a man's experience of army life or of a woman's experience of her first year at college. Much closer to capturing experiences like these is Dewey's description of experience in terms of 'participation and communication' in or with an environment (1980, 22). The environment could be the fields in which people are working, the space before a temple towards which people are processing, the mountaintop on which someone is meditating, or the hall in which an audience is enjoying a show. In keeping with Dewey's description of experience, we should include under the heading of 'musical experience' not just listening to music performed on a stage, but humming worksongs in the fields, drumming or dancing ahead of the priests in a procession and blowing notes on a shakuhachi as one sits alone on the mountain.

It helps to keep this rich notion of musical experience in mind to refer to experiences not only *of*, but *with*, music. At any rate, in the following sections of the chapter, it is upon the wide range of musical experience illustrated in the previous paragraph that my discussion draws. Certainly, if we are to appreciate music's contribution to a sense of convergence with the natural world, then attention must not be confined to the enjoyment yielded by just listening to certain organized sounds.

MATERIALS, MOVEMENT AND SOUND

'Musical instruments', it has been rightly remarked, 'have been part of the human experience in nature from the beginning' (Clifford 2012, ch. 1), and it is clear from many testimonies that making, handling and using instruments have a capacity to strengthen a person's sense of relationship to nature. In chapter 19 of the Daoist classic, *The Zhuangzi*, a maker of bell-stands explains how he enters into the mountain forests to search for a tree with the appropriate 'inborn heavenly nature', in whose wood he can 'see the completed bell-stand' (Zhuangzi 2009, 82). The maker of cedar wood flutes in a Native American style 'look[s] up at a tree thinking about how the wood from this tree will make wonderful music' and in so doing derives a sense of being 'connected with nature' (Clifford 2012, ch. 3).

From testimonies like these, it seems that the feel, look and smell of instruments made from natural materials—let alone the sounds they emit—may invite a heightened awareness of the natural world, of the presence of nature in the products of human craft. But the sense of convergence promoted goes beyond this. The great champion of East Asian traditional and local crafts, Sōetsu Yanagi, reflecting on why 'raw materials' with an 'unspoiled character' are so 'precious' in the making and using of things, musical instruments included, refers to a sense of 'safety' that is felt when we thus remain 'close to nature'. He writes, too, of the feeling that the beauty of such things is experienced as 'a gift, an act of grace' (1989, 194, 215). The American flautist cited above speaks similarly of 'gratefulness' to nature for the musical experiences that nature's raw materials have made possible (Clifford 2012, ch. 8). In such remarks, voice is given to an aspect of unity or intimacy with nature prompted by an experience with music. Gratitude to nature, surely, is a way of coming closer to it.

Musical instruments do not play themselves; they require the intervention of the human body—of hands, lips or feet, for example. But the body's involvement in music is not, of course, confined to this. Whether or not Ezra Pound was right to bemoan the 'atrophy' of music when it is divorced from dance, it is important to recognize that a great deal of music has always been performed in conjunction with dance—and important, for our purposes, to recognize the associations with the natural world that, in traditional societies, were reinforced through dance. As Black Elk explains, in the Lakota Sioux 'Elk Dance' a dancer impersonates—or rather, 'changes into'—an elk, thereby invoking the powers and mysteries represented by this animal (Neihardt 2014, 129). The purpose, it seems, of many Native American dances, and of comparable ones in other cultures, is not magically to influence nature, but to express and confirm a close relationship with creatures, natural processes and environments. The rain-dancer, as Suzanne Langer observes, 'dances *with* the rain, he invites the elements to do their part' (1957, 158). In some cases, it is gratitude to nature that a dance expresses, as when people celebrate the harvest by dancing on a village green—or even, perhaps, when young people dance on a sandy beach by the edge of the sea.

Some dances form part of ritual, but even when rituals do not include dancing, bodily involvement with the music sung, chanted or played remains significant. To understand the effectiveness of music in ritual, remarks one musicologist, it is essential to consider not simply 'the music itself', but the participants' 'bodily experience of [the] music' (Aksnes 1998, 178). Only, for example, those whose bodies the music has relaxed and stilled may be capable of sensing and welcoming the arrival of the forest spirits that the ritual invites into a sacred space.

Bodily experience belongs with musical experience even when no actual bodily movement occurs—as when, for decorum's sake, one remains immobile when listening to a piano sonata in a concert hall. There is movement in music. To be sure, 'musical movement' is a metaphor—for no space has been traversed when the clarinet soars upwards at the start of *Rhapsody in Blue* or the horns plunge ahead in *The Ride of the Valkyries*. But, as Roger Scruton persuasively argues, phenomenologically listeners cannot but 'hear up and down, towards and away, soaring, plunging . . . distance [and] direction' (2009, ch. 4). That music (metaphorically) moves is a candidate for defining its difference from mere sequences of sound, and is also what makes music so suitable a medium for tonal depictions of real movement—of swans flying or waves washing onto the shore. (In Merleau-Ponty's rather grudging admission that music, despite being 'far beyond the world', may depict something, it is 'ebb and flow' and 'turbulence' he mentions [2004, 293.])

Importantly for our purposes, Scruton goes on to remark that we do not just hear movement in music but 'we move with the music, responding to accent and line, as in a dance'—unless, of course, the movement is inhibited, as in a concert hall or church. When incipiently 'dancing with' the music, Scruton continues, we feel as if we are 'invited' to join in with the movement, and to imagine others with whom to move (2009, ch. 4). He has in mind other people, but the point applies as well to the power of musical movements to arouse a sense of joining in with what moves in the natural environment. We feel as if our bodies are swaying with the trees, rocking with the waves or being lifted to the sky. This is why music is good not simply at depicting what moves in nature, but at reminding us—embodied, mobile beings that we are—of our affinity with and immersion in this world of movement.

The part of the body that enables auditory contact with the world is the ear. Sound penetrates the head through the ear. This is not the language anatomists or biologists use, but phenomenologically talk of penetration is revealing. When looking at coloured surfaces, we can usually pinpoint their location and easily distinguish between them and the affective responses we may have to them. It is different in the case of sounds: while one may locate its source, a sound isn't—or isn't only—where the source is, and it is often impossible to separate the sound heard and one's response to it. We don't first hear a sound and only then find it grating, soothing or shattering. The language of penetration helps to register these aspects of auditory experience: Sounds as heard are as much 'in here' as 'out there'—or, rather, the distinction gets no purchase.

It is with considerations like these in mind that the Japanese composer Toru Takemitsu writes that 'sound penetrates me, linking me to the world'. Because the experience of sound enables him to 'merge with the world', he also experiences a 'harmony with nature' which his art strives to accomplish. For not only does nature penetrate him with its sounds, but these sounds are then worked on by the composer before being 'freed' and returned to nature when his pieces are played or sung (2009, 183, 189). Takemitsu, we may be confident, knew the beautiful lines of the great Tang poet, painter and musician, Wang Wei:

> You ask what the principle is for achieving the Way—
> A fisherman's song going deep into the river bank. (in Cooper 2012a, 55)

MUSIC OUTDOORS

We've been concerned, in the preceding section, with aspects of the body's involvement in or with music, and of their bearing on a sense of convergence with nature. In the present section, attention turns to the location of music—in particular, to the significance of its being played or listened to *al fresco*, in or close to natural environments.

Wang Wei enjoyed 'playing the zither . . . among the secluded bamboos', while the legendary Chinese Yellow Emperor not only played his instrument in wild places but, as it were, played along with his surroundings, 'adjusting' to them so as to produce 'unforced and uncontrived' sounds (Zhuangzi 2009, ch. 14). Three thousand years later, and in a more contemporary idiom, the environmental philosopher and clarinettist David Rothenberg urges that it is 'important to get out there and jam, to play with the world and let the world play with us' (Rothenberg and Ulvaeus 2009, 8). These are but a few of the many individuals who find that music played in—and in dialogue with—natural environments deepens a sense of communion with an environment. In the way that one feels close to another person to whose moods one spontaneously responds, so the musician feels an intimacy with creatures and places whose moods elicit an immediate and effortless response—and an intimacy too with creatures that in turn respond to the music—like the fish that Thoreau 'charmed' with his flute on warm evenings (1886, 173).

It is not only individuals but whole communities for whom this dialogue is important, in the context of ritual, for example, or tribal celebration. The music played and sung by the Kaluli people of Papua New Guinea, writes the ethnomusicologist Steven Feld, is 'inspired by, modeled upon, and performed with . . . environmental sounds' (2009, 195). Louis Sarno's remarkable chronicle of the musical life of the Ba-Benjellé people of the Central African Republic describes how the women, when gathering fruit in the jungle, would burst into songs that echoed the voices of the birds and insects around them—just one example of the many ways in which this people's music confirms that the forest is their home (1988, ch. 4).

Such dialogue between music and nature is of interest, not only to anthropologists but to contemporary composers and sound-artists for whom it is a self-conscious theme and ambition. The dialogue, for many of these artists, is not an end in itself but a contribution to reinforcing a sense of humanity's convergence with the natural world. For the Canadian pioneer of soundscape art, R. Murray Schafer, with the retrieval of the 'original context of music'—'the outdoor environment'—there is the prospect of 'music, environmental sounds and silence' being 'woven together' into a 'new consciousness' of the relationship between humanity and nature (2009, 60, 68). For another sound-artist, the outdoor performance of works that are themselves 'hybrids' of electronic and ambient sounds can be a 'prime integrating factor', helping to bring to mind a 'wholeness . . . of which we are intrinsic part' (Dunn 2009, 97, 99).

It is not, of course, just avant-garde sound-artists who have wanted to blend human music with natural sounds by way of expressing a feeling of unity with nature. Beethoven, who was sympathetic to the idea that God may be found 'in the whole of nature', is famously reputed to have remarked that when composing a scene in his *Pastoral* Symphony, yellow-hammers, nightingales and other birds 'composed with me' (Robbins Landon 1970, 129). For Olivier Messiaen, the reason he 'has taken bird songs as a model' for many of his pieces is precisely to display a 'union' between 'the sphere of men' and that of birds and other creatures (Griffiths 2012, ch. 10).

Not all composers with an ambition to blend music with natural environments have wanted to achieve this by incorporating the sounds of nature, such as birdsong, into their works. The excessively ambitious Russian composer Alexander Scriabin never completed his enormous work, Mysterium, that he intended to be performed over the course of a whole week high up in the Himalayas. (It would have consoled him to know that an enthusiast, with a CD player, would one day climb Mount Melungtse and from its summit boom out a piece based on Scriabin's sketch for his work.) Less grandiosely than Scriabin, Claude Debussy also dreamt of writing 'a kind of music composed especially for the open air . . . which would sport and skim among the tree-tops in the sunshine and fresh air' (1927, 32). As these words suggest, Debussy's point was not that—on a balmy evening with the smell of roses wafting about—people would enjoy listening more than they would in a stuffy concert hall. Rather, the sounds of the voice or orchestra would enter into dialogue with the breeze and the movements of the trees. The 'greater harmony with natural scenery' achieved by outdoor performance would enhance the experiences of music and scenery alike—or, better, it would conjure an experience impossible to bisect into these two halves.

The impossibility on such an occasion of separating the musical experience from that of the surrounding environment is a clue to the more pervasive experiential union or convergence between music and nature to which I now turn.

'MUTUAL INFLECTIONS'

According to the Confucian classic, the *Liji* (*Record of Rites*), music and the rites are created in response to heaven and earth and, when 'fully realized', help 'heaven and earth to function in perfect order' (de Bary 1960, 168). It is tempting to regard this dramatic claim—and comparable ones in other ancient traditions—about harmony between human music and the universe as gesturing towards a deep insight of phenomenology rather than as a truth about the cosmos. Doubtless, there are any number of two-way causal interactions—including several noted earlier in this chapter—between musical performance and events in the natural world. But it is a truth about the relation between experience of music and experience of the natural world that I want to articulate. It is one hinted at in the words of Steven Feld: In the musical engagement of the Kaluli with nature he discerns 'the naturalization of music as an ecologically modeled system and the culturalization of nature as an aesthetic system' (2009, 199).

These words register that how people experience the natural world is partly dependent on their musical culture and, conversely, that their experience of music is partly dependent on their engagement with the natural world. We have here a 'codependence' between the spheres of experience—a special case of a larger codependence between culture and nature. Cultural practices and perceptions are always already shaped by how nature is encountered, just as ways of encountering nature are always already shaped by cultural traditions (see Cooper 2012b). Given the context of his remark that 'it is impossible to say that nature ends here and that man or expression starts here'—a discussion of painting's relation to Being—this thesis of codependence is what Merleau-Ponty may have intended.

That knowledge of nature affects how music is heard is, at one level, too obvious be worth dwelling on. Clearly a great deal of music is recognized as mimicking natural sounds and depicting natural places. But there is a different and less facile way in which experience of nature inflects the music we hear, irrespective of whether the music is intended to depict nature. Gabriel Marcel attributes the powerful experience afforded by a work of Debussy's to a 'way in which nature is present in this work', the way that its composer has himself been 'invested and . . . penetrated by the mysterious power of nature' (2005, 130, 142). What Marcel hears in Debussy, others also hear in the music of different composers. In a rare remark on music, during a memorial address to a composer from his own birthplace of Messkirch, Heidegger refers to 'the life-giving powers of his Heuberg homeland' that are present in Conradin Kreutzer's music (1969, 56). Works of Sibelius have been described as seemingly 'grown out' of nature, and as 'standing', not *for*, but '*in for*', nature (Grimley 2011a, chs. 7 and 13). Gustav Mahler described his Third Symphony as a work in which 'the whole of nature finds a voice', and of nature, rather than himself, having composed his music (Johnson 2012, ch. 4).

Importantly, it is not simply the composer, such as Mahler, whose musical experience is invested by nature—for so is ours, the listeners'. As Heidegger indicates, it is not Kreutzer alone, but the people of the region, whose experience of the music is formed by 'the life-giving powers' of their homeland. Listeners can only speak, with understanding, of works as 'grown out' of or 'penetrated by' nature if they appreciate how their own experience of nature inflects their reception of the music.

When, in defence of the thesis that 'Nature Is an Imitation of Art', Oscar Wilde remarks that there were no fogs before Turner and the Impressionists painted them, he was joking—but with serious intent (1966, 985f). How nature is perceived cannot but reflect cultural traditions in which it has been portrayed and written about. The point is similar to Heidegger's when explaining how the Greeks' experience of the earth was enabled by a culture—its buildings, sports, drama and so on. Wilde and Heidegger make no mention of music, but this too plays its part in what Feld called 'the culturalization of nature'. Consider, for example, how perceptions of different kinds of animal—as cute, fearsome or whatever—have been influenced, from *Bambi* onwards, by the clichés of movie scores. Equally, the expectations and selective attention of people who travel through the American West have, in part, been moulded by listening to a large body of music that, often in the form of film scores, has evoked the prairies, canyons and deserts. If Sibelius's music, as just noted, is 'grown out' of 'the northern forest realm' of which so much of it is a 'sound-portrait', then the composer repays his debt—for the music, as so many visitors to this realm testify, serves to 'shape [a] view of the world', to invest it even with notions of 'finality and death' (Grimley 2011b, 395f).

Here, then, are a few indications of what one musicologist has usefully labelled 'the mutual inflections of music and place' (Watkins 2011, 408). The pervasiveness of these inflections, she writes, give the lie to 'the idea of musical autonomy', to a conception of musical tradition developing in splendid isolation from human beings' engagement with nature. She might have added that it gives the lie, too, to a conception of the autonomy of nature, to the thought that encounters with nature can be raw and immediate, unshaped by our cultural practices.

My indications were of modern Western musical experience, but the truth of codependence is, if anything, more salient in the case of traditional societies—like those of the Lakota Sioux, the Kaluli or the Ba-Benjellé—where music was and is largely made, played and sung in contexts involving a self-conscious interplay with a natural environments. But though it may be less obvious, this 'transformative play of nature and culture', as Feld calls it (2009, 200), is there to be discerned as well in modern Western societies.

CODA: MUSIC, NATURE AND CONVERGENCE

Codependence is a form of convergence or unity. When, for example, the lives of two people shape one another, these lives converge, and when they recognize this, they share a sense of convergence. Likewise, to appreciate that a musical culture and a way of engaging with nature are codependent is to have a sense of their convergence and intimacy. Here, in effect, we have an illustration, or dimension, of the idea of 'the human world'—of the world as a sphere of significance that cannot be even notionally isolated from human practice, perception and purpose.

But exactly what, someone might ask, *is* this sense of convergence between humanity and nature of which appreciation of a codependence between music and environment is one dimension? The best answer to this question, I suggest, is simply to recall the various aspects of the experience of music in relation to the natural world that I have discussed—'gratitude' to nature, 'penetration' by natural sounds, 'dialogue' between music and environment, bodily engagement with nature through music and the impossibility of finally separating the contributions of culture and nature. It is aspects of experience like these that constitute a sense of convergence or unity with nature.

We should not, it seems to me, be looking for a single eureka recognition of some alleged metaphysical truth—the identity of subject and object, say, or the nonexistence of individual selves—of the sort that some eco-philosophers hope to establish (see Cooper 2012a, ch. 13). For one thing, it is unclear how such a recognition could play out in a person's life and be the remedy for an estrangement from nature that cultivation of a sense of convergence is intended to be. This sense must inform and show up in a person's life if it is to further an aspiration to consonance with the world in which this life takes place. This is why it needs to be sought in various achievable and often simple forms of reflective experience, like those that become salient when we reflect upon the engagement of music with the natural world.

FOURTEEN

Phenomenological Aesthetics of Landscape and Beauty

Guðbjörg Rannveig Jóhannesdóttir

When Husserl criticized the European or Western scientific frame of mind for changing the world of perceived nature into a mathematical world, he was addressing a deep-rooted problem that has had an enormous influence on how Western societies frame their relationship with the natural world. In this mathematical world that Husserl described, everything that cannot be submitted under a scientific, objective calculation and is therefore 'merely' subjective, must be eliminated and declared as worthless, inexact and insufficient (Husserl 1979/1936). This frame of mind has strongly influenced debates on environmental values, especially when it comes to decision-making processes where the aim is always to find ways to measure environmental values objectively, preferably with the methods of the natural sciences or economic theory. This has led to the neglect of environmental values that cannot be measured easily, such as the aesthetic value of landscape.

The idea that the beauty or the aesthetic value of a landscape cannot be measured and thus taken into account in decision making as it is subjective and relative—a personal emotion—is rooted in the common account of beauty as being either subjective: 'in the eye of the beholder', or objective: in the physical qualities of objects. This account of beauty has played a central role in shaping our understanding of the concepts of landscape and the aesthetic. Even though efforts have been made within philosophical aesthetics to understand the nuances between the subjective and the objective and to secure an intersubjective ground for aesthetic judgments, such efforts remain largely unacknowledged in common discourses where the line between the objective and subjective are sharply drawn. This is in line with the polarization of Western thought through the ages, where our focus on the differences between the dualistic poles on each end prevents us from seeing the 'in-between'.

This state of affairs has influenced debates on environmental controversies surrounding land and land-use in Iceland. The word *beauty* seems too simple and superficial to describe the deep aesthetic experiences (not limited to the visual) that people have of the dramatic landscapes of the Icelandic highland; and in the public debate and environmental decision making in Iceland, beauty has been judged as too linked to individual taste to form a basis for any meaningful evaluation. In order to reclaim beauty as a way to assess the value of landscape, we need to understand the concept of beauty differently. We need a concept of beauty that covers more than primarily visual qualities and we need accounts of subjective perspectives that are relative but yet have general features.

The concept of beauty that I will examine here in order to shed light on the landscape concept is situated on a different level of perception and experience than provided by approaches that are determined by the subjective-objective divide. It is not a concept that generates some kind of 'standards' of beauty for assessment such as depicted with terms from aesthetics of the work of art like equilibrium, balance, proportion, etc., and it is not a concept that is limited to describing in line with aesthetics of perception, feelings of pleasure, objective qualities or standards of taste that are determined by time and place. Rather, the concept elaborated on the basis of aesthetic experiences of landscape generates an understanding of what we mean when we describe something as beautiful. This concept of beauty is a relational concept that describes the moment where your senses are captured, when your whole being is immersed in what you perceive so that something is revealed to you: the relations between yourself and the other. Beauty is the moment where the boundaries between the perceiver and the perceived disappear. Beauty felt in a natural landscape is an experience of relationality and as such it is an experience central to the human condition. An important part of knowing what it means to be a relational human being is thus knowing beauty.

Through examining the concepts of beauty and landscape from the perspectives of phenomenological and environmental aesthetics, my aim is to shed light on questions concerning the contribution of phenomenological aesthetics to our understanding of natural beauty as well as shedding light on questions concerning the role nature and beauty play in the constitution of the ethical self. I will also reflect on some practical issues concerning the issue of using science to discuss the environment, as the aim of the research this chapter is based on is to create a space in environmental decision making for values that cannot be measured 'objectively' with scientific methods and have therefore been neglected and deemed immeasurable, subjective and relative.

LANDSCAPE AND BEAUTY

The concept of landscape has theoretically many layers, but one of its core philosophical aspects is captured in Joachim Ritter's definition of landscape as an environment perceived aesthetically (Ritter 1989). However, this philosophical concept of landscape has not received much attention in contemporary environmental debates, except in a limited way that reduces its full aesthetic aspect to a much more simple visual perception of scenic qualities. Thus, there is a lack of attention to the core of the aesthetic aspect of landscape as an environment perceived aesthetically in a broader way, through all the senses, by embodied, relational beings.[1] The reason for this neglect is the manner in which the landscape concept has been understood and framed in the context of landscape assessment in environmental decision making.[2] The framing of landscape in terms of the subject-object dualism is intricately related to the difficult status of beauty within landscape studies. The issue of beauty is obviously paramount in descriptions of perceptions of landscapes, yet beauty is for the most part deemed to be merely subjective. This has resulted in its expulsion from landscape assessment.

The phenomenological concept of beauty that this chapter elaborates discloses a sense of relation felt by an environmental self; it provides a possibility of experiencing a basic feature of our condition as embodied beings. The sensory immersion that is involved in the experience of beauty is a source of 'sensory knowledge'. This is an experience that stirs our emotions and allows us to perceive the world in a manner that reveals to us our relations with the environment that are in a certain ontological way deeply ingrained and at the same time conditioned by actual circumstances. The phenomenological concept of the aesthetically perceiving subject is a concept of the self as a relational entity that cannot be abstracted from its surroundings. The self is always already in relation with itself, reflecting on itself and its place in the world; it is always already in relation with the other, as it sees and is seen by the other; and it is always already in relation with its environment. This last notion of the self as an 'environmental self' is crucial to an understanding of aesthetic perception of landscape.[3] The phenomenological idea of the relational self reveals an understanding of the human being that acknowledges the central role of our embodied being in shaping the values and meaning that we find in the world. This idea reveals how the values we associate with natural environments emerge from the intertwining of subject and object in the experience of nature; thus they should be framed in terms of the relations that are created in these experiences, rather than in terms of measurable qualities of separated objects.

The systems that Western societies have created around environmental decision making are not able to take this relational understanding of the self in an environment into account. These systems are based on the mathematical world view that Husserl criticized—a world view where we have come to rely so much on numbers, mathematics and calculation in order to represent our knowledge of the environment and the values we associate with it that we have come to believe that things that are not quantifiable are not worth identifying. Or perhaps we don't even believe that they are not worth identifying, but rather we simply don't know how to deal with the immeasurable aspects of human existence and therefore we get confused when we are confronted with them.

To exemplify this, let me tell a story from a meeting where landscape professionals in Iceland came together to discuss ways of dealing with the value of landscape in decision making, methods of finding out what landscape is and what it is worth. At the meeting there were many landscape architects, planners and professionals who all had very good technological methods to deal with landscape. Everything that could be quantified was quantified and made visual through maps and graphs and numbers. But one of these professionals had a significant complaint: 'We all have very good methods of quantifying and visualizing the landscape's objective qualities—and this is an important part of landscape evaluation—but we don't know how to deal with what landscape really is: what is in here', he said as he pointed to his head. And the others agreed with him. Thus the problem was not that they didn't acknowledge the existence of the aspect of landscape that cannot be quantified, they just didn't know how to deal with it.[4] How do you measure the immeasurable?

This confusion or crisis that arises when we are confronted with the qualitative dimensions of existence has its roots in the divide that we have been taught to see between the subjective and objective. The reason these professionals have not found a way to deal with the 'subjective' aspect of the landscape concept is that this aspect is (and has been for a long time in the Icelandic language)[5] connected in their minds to the concept of beauty. Landscape is understood in terms of experiencing beauty in nature, and beauty is understood as being purely subjective and thus also relative, which leads to the idea that there is no way of taking these experiences of beauty into account because there is no way of coming to an objective conclusion about something that is inherently subjective.[6]

From this we can see that the particular qualitative dimension of human existence we choose to call the aesthetic value of landscape has been seen as very difficult to deal with; the result is that the values associated with this dimension have been ignored in decision making. But what can we do to change this? How can we deal with the qualitative dimensions of human existence? I will argue here that the solution to this problem must involve moving beyond the understanding of beauty and landscape in terms of a divide between the subjective and the objective.

THE QUEST FOR OBJECTIVITY IN AESTHETICS

Even though discussions of the relation between the subjective and objective are commonplace in philosophical aesthetics from Kant and Hume to current debates, the knowledge gained through these discussions has not managed to undermine the striving for objectivity in assessment of landscape that is driven by technocratic demands for scientific accuracy. If there was awareness of the untenability of strict objectivity, it would be accepted that the aesthetic value of landscape should be framed in terms of the relation between the physical qualities of the land and the aesthetic experience of it, rather than only in terms of its physical qualities, as has often been the case. Since the aesthetic experience of landscape is assumed to be too subjective and relative to be taken into account in evaluations and decision making, this aspect of the value of landscape has simply been left out. Thus, dealing with the value of natural landscapes has been reduced to dealing with objective physical features of the land that can be shown to be rare, diverse or unique.

These kinds of standards and especially the 'objective' methods derived from them are at a closer look arguably not even as objective as they are presented to be. My point here is a phenomenological one, that we are dealing with meanings that are given based on certain experiences as well as diverse interests when discussing landscape values. Thus there is always a 'subjective' decision made somewhere on the way to an 'objective' evaluation. Why is it so hard to admit that our 'objective' standards are not strictly objective after all, and that neither are our allegedly subjective standards completely subjective? Do we necessarily have to rely on the idea that there exists some kind of strict objectivity or subjectivity? (Berleant 2010b; Harding 1991).

These remnants of the dualistic schemes that have permeated Western philosophy have tied philosophical aesthetics, landscape studies, other academic disciplines and environmental decision-making processes up in knots that need to be untied. I agree with Arnold Berleant (2010b) that in order to untie these knots we need to look to all the theories and insights that have already started to move beyond the subject-object dichotomy in the last century, and for my purposes, Merleau-Ponty's thought is a very useful source to look for guidance. In the next sections I will therefore explore what an approach to aesthetics inspired by Merleau-Ponty's philosophy looks like and how it changes the way we think about and frame the aesthetic value of landscape.

Our understanding of the concept of beauty directs our understanding of the aesthetic, so if we are to get beyond the dualistic understanding of the aesthetic and the value of landscape, we need to rethink the concept of beauty. There is however a tendency to avoid the concept of beauty, precisely because it is assumed that it leads to subjectivism about aesthetic value. With the aim of getting beyond the subject-object divide in the understanding of landscape, the tendency within landscape studies has been to emphasize the role of perceptual experience in creating the meaning of landscape without referring to the concept of beauty or even the aesthetic aspect of this perceptual experience (Ingold 1993; Casey 2001; Massey 2006; Wiley 2007; Dewsbury and Cloke 2009).

I argue first that the perceptual, emotional and bodily experience of landscape *is* the aesthetic experience of landscape and secondly that the concept of beauty should play a key role in understanding the aesthetic experience of landscape. The understanding of beauty that I will propose does indeed provide a basis for shedding light on the intertwining of subject and object that characterizes the perceptual, emotional, bodily or, simply, aesthetic, experience of landscape. As I will explain, one of the main characteristics of beauty is the rapt attentiveness that is given to sensuous perception in the experience of beauty and thus understanding this aspect of beauty should be the starting point for understanding the aesthetic (perceptual, emotional) experience of landscape.

Within environmental aesthetics, the tendency has also been to neglect the concept of beauty as such. Even though natural beauty is of course the main topic of environmental aesthetics or more specifically the aesthetics of nature, the concept of beauty itself has not been central. Rather terms such as aesthetic appreciation, aesthetic perception or aesthetic experience have been in the leading role in theories within environmental aesthetics. Perhaps many authors find it unnecessary to discuss beauty specifically as they see it as being an inherent part of discussing aesthetic experience or appreciation; perhaps they are using aesthetic experience as a synonym for beauty. Whatever the reason, this is usually not something that is considered to need any specific explanation. But I in contrast think that it does need to be explained and that there is an important reason behind the low profile of the concept of beauty within both environmental aesthetics and landscape studies. As will be discussed in the following section, the concept of beauty has come to be seen, largely, as uninteresting and trivial both in art and philosophy in the last century or so. It has been devalued and neglected; this history of the concept is arguably an important part of the reason that beauty has not been treated as central in environmental aesthetics, landscape studies and other fields that deal with perception and the aesthetic. However, the main reason behind this devaluing and neglect of beauty has to do with the narrowing of the concept of beauty that especially modern art contributed to. But instead of giving up on the concept for this reason, I argue that one should rather just leave this narrow understanding behind and revisit the concept of beauty from a phenomenological perspective.

RECLAIMING THE CONCEPT OF BEAUTY

As mentioned above, the concept of beauty has been central to our philosophical understanding of what the aesthetic is. In everyday language the words *beauty* and *beautiful* are more or less unaffected by philosophical critique of these terms, yet they are clearly quite elusive and hard to define. The word *beauty* is so commonly used that every person probably claims to know what it means, until she is asked to define it. Why is this concept so complicated in its simplicity? Is it because the experience of beauty which seems so simple—to be fascinated by the red colour and sweet smell of a rose, or the golden rays of sunset by a glacier—affects our consciousness in a complex and deep manner? Is the experience of beauty one of the most important pillars of human life? If so, perhaps it should not be so surprising that defining the concept of beauty has proved to be so difficult. We have here a similar situation as when dealing with god and the divine: beauty and the aesthetic are names that we have given to certain distinctly human experiences that play an important part in creating the core of what it means to be human, but at the same time these are experiences that are difficult to express in clear and concise terms.

As is common with elusive concepts, the concept of beauty has been understood and misunderstood in many different ways through history. After it almost disappeared from twentieth-century aesthetics,[7] many scholars have tried to revive the concept of beauty and bring it back into focus in the last decades (Mothersill 1984; Kirwan 1999; Scarry 2001; Danto 2003; Johnson 2009; Brand 2000, 2013; McMahon 2011; Heller 2012). According to the common account of beauty that I am dealing with here, beauty refers chiefly to the form and appearance of an object, and the experience of beauty should be distanced and disinterested. A phenomenological account of beauty is very different. For Galen A. Johnson, the concept of beauty not only refers to form and other sensuous qualities (colour, sound, smell), but it also refers to a certain relation between subject and object; in this relation there is not distance, but rather a certain closeness and openness between subject and object (Johnson 2009, 6). When we experience beauty we open ourselves up to the object in a unique way. Phenomenology of the aesthetic is thus necessarily based on an aesthetic desire for the beautiful.

THE DESIRE FOR BEAUTY: OPENING UP TO THE OTHER

In order to understand the opening up to the other that occurs in the aesthetic experience of beauty in landscape, it is useful to examine how Johnson uses Merleau-Ponty's idea of an aesthetic desire. For Merleau-Ponty, aesthetic desire does not refer to the idea of a negative desire to own, control or use the object that is often referred to in Western thought from Plato to Sartre. Rather, his idea of aesthetic desire connects with a positive notion of desire that entails opening oneself up towards the object as it is; it involves an openness towards the other to receive the meaning that the other expresses. The core of the phenomenological account of beauty involves opening oneself up to the other[8] and listen, see and feel who the other really is instead of projecting our own concepts and ideas on the object of beauty and thereby striving towards controlling and owning it.

In this context Johnson refers to Heidegger's ideas of the experience of beauty, but Heidegger claimed that with the idea of disinterestedness Kant did not mean that the experience of beauty should be absolutely distant and free from longing but rather that such experience involves a certain attempt to let things be as they are. This 'letting-be' (*sein lassen*) involves being open towards the object and relating to it, and this is the core of beauty: 'for the first time the object comes to the fore as pure object and that such coming forward into appearance is the beautiful. The word "beautiful" means appearing in the radiance of such coming to the fore' (Johnson 2009, 160).

Beauty attracts us; it draws our consciousness towards itself and creates a desire for repetition and for sustaining and sharing beauty with others. Who hasn't had the experience of finding something beautiful and feeling the need to point it out to the next person? Or feeling the need to take a photo, draw a picture, make music, write a poem or simply to take it in for a long time—a need to capture the moment of beauty and create new beauty, continue it in some manner? Beauty nourishes the one who experiences it and this nourishment encourages her to reciprocate: to protect and nourish this beauty so it may continue to nourish others. If our understanding of the concept of beauty is built on this type of desire of openness instead of the negative desire described above, a different picture appears of the experience of beauty and how it affects human life and awareness. This type of experience gives an opportunity to contemplate and dwell on the object.

AESTHETIC EXPERIENCE OF LANDSCAPE AS DECENTRING OF THE SELF

In his essay on walking up the mountain, Francesco Petrarca has to remind himself not to revel too much in the beauty of the landscape because it may, according to the Christian mores of mediaeval times, distract him from thinking of God (Petrarca 2006). What is implied here among other things is an acknowledgement of the fact that a beautiful landscape can make one forget not only the creator, but also forget oneself in a certain way. This is an important reason for seeking such experiences; they give us a sense of relief from mundane thoughts and everyday worries.

An essential part of this experience of opening oneself up for and dwelling in the moment of beauty is a 'radical decentring' of the experiencing subject. When this radical decentring or un-selfing takes place we gain a different kind of access to the world, to what we perceive. Instead of 'the I' being in the centre, the being of what is perceived, independent of all my interests and longings, becomes the centre of attention. There is a sense of unrestraint in such an experience that gives space for disclosure to happen. The result of this is that something is opened up for us so that we have no better way of describing it than to say 'oh, how beautiful!' This allows us to understand more clearly Heidegger's account of disinterestedness as 'letting-be' as openness of the subject towards the world.

What I refer to here as beauty begins in simple perception but when everything is counted, it is much more than that (and perhaps 'simple perception' is also something much more than we usually think it is). The experience of beauty gives us an important lesson in putting our own self to the side. Such an experience entails transcending our narrow interests, altering our preconceived ideas and perceiving things in some new way, and most likely not as we expect them to be. There is hence a transformative moment to the experience of beauty. One's own ideas are modified or enriched when seeing something new or something we already know in a different light—and the important thing is that this radical decentring that takes place in the experience of beauty[9] leads to pleasure, so an individual that has once been captured by beauty is likely to look for opportunities to repeat this experience. Thus the longing for beauty becomes a capability that is cultivated and inspired by each individual's experience.

PHENOMENOLOGICAL AESTHETICS OF LANDSCAPES

If the core of our understanding of beauty becomes relational, how does our understanding of the aesthetic change? According to the phenomenological interpretations of beauty that have been discussed here, the concept of the beautiful does not refer either to physical objective qualities or to subjective experiences of a detached mind. They refer to the most immediate and direct relations we have with the world, and to ourselves via the world. If we apply this relational thought to the aesthetic, we reach a point where the original meaning of the Greek word *aisthesis* becomes central. The sphere of *aisthesis*, or the field of the sensuous, 'marks the realm in which aesthetics as a traditional discipline meets phenomenology—a meeting in which aesthetics has changed and phenomenology has expanded its framework' (Embree and Sepp 2010, xvi).

The connection between aesthetics and phenomenology has often been pointed out as these disciplines share two of their central ideas: the emphasis on 'bracketing' all pre-conceived ideas and opinions, and the emphasis on perception.

This connection unveils how an aesthetic approach to reality can be seen as a form of the phenomenological reduction. In the moment we experience beauty, all our senses are so engaged with the object that all our preconceived opinions and ideas fall into the background; our direct perception becomes the only focus and thus we can see, hear, feel who/what the other is instead of throwing our own ideas, interests and concepts over it. In Arnold Berleant's view, aesthetic perception is a perceptual experience that needs to be recognized as 'the clearest measure by which to assess the values that emerge. Aesthetic perception may indeed stand as the touchstone of human values' (Berleant 2010a, 30). This means that this type of human experience gives us access to perceive the world and ourselves as part of an interdependent system. It is one thing to know things scientifically through scientific ideas and concepts, but another thing is to perceive, to know through the senses as embodied beings. This knowledge of ourselves as beings that are related to the environments must also be accounted for in our systems of decision making by listening to the stories we tell of our aesthetic experiences of relating to the landscape.

Understanding landscape, beauty and the aesthetic from these phenomenological perspectives, we can see that there is much more involved in the aesthetic value of landscape than a distant appreciation of physical 'objective' qualities such as form and colour. If beauty is about relation and opening up to the other, the aesthetic realm can be found when we open up to experience objects or environments 'in a manner that is fully aware of [their] perceptual richness and in which immediate, qualitative perception dominates' (Berleant 2007, 3). The aesthetic experience of beauty thus can become a deep experience that not only gives us pleasure but also helps us better understand our place in the world and our relations to the environments we dwell in.

Reflecting on the aesthetic experience of landscape helps us question some of the most deeply held assumptions that we have about the self, and about the ways we find meaning and value in the world. Landscape is a multilayered concept that includes both the visible and the invisible relationships which emerge in people's actions, movements, speech, thoughts, imaginations and narratives that are intertwined with the visual. If landscape is viewed from this relational perspective, we can see how it adds to our usage of the concepts of nature and culture and bridges the gap we often create between them. Landscape involves a comprehensive way of looking at the reality of places and spaces. Instead of dividing reality into different boxes and viewing nature in one box and culture in another, the landscape concept invites us to look at the whole picture:

how the meaning of landscape emerges in the relation between subject and object, between the human and the land. Taking the value of landscape into account in decision making thus requires creating a space where this relation can be expressed, shared and deliberated; it requires a space for increased public participation.

We have been using the concepts and methods of the natural and mathematical sciences to evaluate the natural world for the last decades, determining the value of biodiversity and ecosystem services, and hoping that these methods will create a solid grounding for good decision making. Although such methods can be helpful, they will never be able to capture the environmental values that are most important. These methods involve approaching the environment as a list of separated objects, where the relations between ourselves and these objects, beings and environments are neglected. Thus the systems that we currently use to deal with environmental decision making are too reductionist; they are not able to account for how relations, time and history create the meaning and value we find in both natural and cultural landscapes.

We find value in the story of the landscape, whether it is the story of human actions or the biological or geological story of the actions of natural forces. The landscape concept that I have discussed here can become a tool in decision making that reveals these stories and relations, and thus provides a fuller account of environmental values than the calculative methods of economics and the natural sciences. Instead of itemizing and separating valuable objects, the landscape concept allows for a more holistic approach; landscape includes all the different layers of stories that contribute to our aesthetic perception of it. Landscapes in Iceland have many stories to tell. First, there is the geological story that can be told by the geologists and by the people who have experienced the influence of geological events like volcanic eruptions, melting glaciers and earthquakes. Secondly, there it the vegetation side and the animal side of the story told by the biologist or the farmer. Thirdly, the archaeological and anthropological stories of human settlement and development in the landscape; and finally the stories of the people who live and dwell in the land and relate to it through their experiences, memories, stories, writings, literature and art. These stories must all be heard in decision making[10] and they can be heard by taking the aesthetic value of landscape into account. When we perceive the landscape aesthetically we are perceiving all these narratives as they are combined into one interconnected whole, and this gives us the opportunity of having the experience of being whole, and of being interconnected with this whole (Skúlason 2005, 2008; Thorgeirsdóttir 2005, 2010). As we perceive the landscape through our aesthetic embodied attention, we experience relationality. We experience being related to the earth as we perceive the story of its creative forces through lava fields and glaciers; we experience being related to the humans that lived in the landscape before us as we perceive their story in old ruins; and we experience being related to the vegetation and the animals around us as we watch them live and grow. In this way we can find meaning in the landscape beyond its meaning as a resource or as a view out the car window.

Taking the value of landscape into account means changing the way we frame the human-nature relationship in our systems of decision making. Instead of seeing ourselves as separated subjects that measure, count and control the objects of nature, the landscape concept allows us to see how there is no separation between humans and nature, subject and object. It allows us to express and deliberate the human-nature relations that are revealed in aesthetic experiences of landscape. These are meaningful relations (Holland and O'Neill 1994) that are based on a certain kind of love of nature and feeling connected to it. Meaningful relations are like relations of love and friendship, based on the deep aesthetic of 'letting-be' and perceiving what/who one loves or relates to as an independent other. Finding beauty in the other in this deep and rich way which allows one to open up to the other is thus a key to forming meaningful relations. If our aim is to transform our societies and question deeply held values and assumptions about our place in the natural world, these are the types of human-nature relations that we need to create a space for in our deliberations and decision-making processes.

NOTES

1. This type of landscape concept has also received very limited attention within Anglophone philosophical thought. This has even been the case within environmental aesthetics where the concept of landscape has been commonly associated with scenery and visual approaches to nature. This is evident in criticisms made by environmental aestheticians of the 'landscape model' for aesthetic appreciation of nature which is described as a model that emphasizes scenic, visual qualities of nature but neglects other aspects of the environment and our aesthetic appreciation of it (Carlson 2000, 45–47). The notion of landscape as related first and foremost to scenic and visual qualities is thus taken for granted within environmental aesthetics, and attempts to expand the landscape concept to include more than the visual, or critically reflect on its meaning, are lacking.

2. In environmental discussions, the concept of landscape is commonly understood and framed in terms of the subject-object dualism that has characterized Western thought for a long time. This way of framing the concept of landscape has led to its weak position when it comes to environmental decision making in Iceland. This is also evident in Brady's comment about the UK and U.S. contexts: 'When aesthetic value is given a formal place in conservation strategy and decision-making, there is a tendency to value scenic and formal qualities over other kinds of aesthetic features. Landscape conservation strategy in Britain is directed to a great extent by scenic critera. . . . The scenic approach appears to be widespread in North America too' (Brady 2003, 225). This 'scenic model' is based on the dualistic view of the distanced subject viewing the objective formal qualities of the landscape, and this approach is limited and superficial; it does not allow for a deeper, more inclusive understanding of the range of aesthetic values that are at stake in the aesthetic appreciation of landscape. See also Brady 2006.

3. For a discussion of the environmental self, as one aspect of the multiple self, see Zahavi 2012.

4. This confusion and conflicts about the use of the landscape concept are also discussed in Edda R. H. Waage's research. See Waage and Benediktsson 2010.

5. Waage's research into the origin and everyday use of the word *landslag* in Icelandic confirms this. See. Waage 2010; 2012.

6. This view can be seen in a report on The Nature Conservation Strategy 2004–2008 from the Environment Agency of Iceland (Harðardóttir 2003, 34), as well as in the above-mentioned qualitative case study of the preparation for the Icelandic Nature Conservation Strategy 2004–2008, where contributors to the strategy were interviewed (Waage and Benediktsson 2010, 18).

7. As Kirwan points out: 'When the word beauty is used in aesthetics—as it sometimes is, in the forewords, conclusions and even titles and subtitles of books on the philosophy of art—it is used as a mere rhetorical flourish: a word either without any meaning whatsoever, or given a meaning (for example, as a blanket term for the merits that art may possess) that is quite at odds with that normal usage which constitutes its meaning in the world beyond. . . . "Beauty" quietly slips away from the serious business of philosophizing about art—as, indeed, it should' (Kirwan 1999, viii).

8. The concept of *the other* has been a key concept in Continental philosophy and is often used to describe how our definition of the 'other' is a part of what constitutes the self, and to understand how societies and groups exclude 'others' whom they want to subordinate. In my use of this concept I focus on how it describes the inherent relationality of the human subject that is always already in relation with the other, and how aesthetic experiences can provide the possibility of opening up to the other in a unique way.

9. Not all aesthetic experiences of beauty are equally powerful in the way that they provide a possibility of this type of decentring of the self. However, they all do hold the potential to draw the perceiver's attention to her immediate sensuous perception in a manner that allows her to open up to the other in this unique way. In cases of extreme and sublime beauty, for example when standing up against a glacier, the self is 'thrown' into a relation with the other in a dramatic manner, while in cases of simpler beauty, for example of a flower, this relation is revealed in subtler ways.

10. Martin Drenthen (2013) argues in a similar vein for the importance of narrative in environmental decision making.

References

Abram, David. 1996. *The Spell of the Sensuous: Perception and Language in a More-Than-Human World*. New York: Vintage.

———. 2010. *Becoming Animal: An Earthly Cosmology*. New York: Random House.

Agyeman, J., P. Cole, R. Haluza-Delay, and P. O'Riley. 2010. *Speaking for Ourselves: Environmental Justice in Canada*. Vancouver: University of British Columbia Press.

Aksnes, Hallgjerd. 1998. "Music and its Resonating Body." In *Proceedings of the 5th International Conference on Musical Signification*, 173–82. Bologna: CLUEB.

Alaimo, Stacy. 2000. *Undomesticated Ground: Recasting Nature as Feminist Space*. Ithaca: Cornell University Press.

Alexander, Christopher, Sara Ishikawa, and Murray Silverstein. 1977. *A Pattern Language*. Oxford: Oxford University Press, 1977.

Appadurai, Arjun. 2004. "The Capacity to Aspire: Culture and the Terms of Recognition." In *Culture and Public Action*, edited by Vijayendra Rao and Michael Walton, 59–84. Stanford: Stanford University Press.

Arendt, Hannah. 1998. *The Human Condition*. Chicago: University of Chicago Press.

Aristotle. 1995. *Politics*. Translated by Ernest Barker, revised with an introduction and notes by R. F. Stanley. New York: Oxford University Press.

Arquette, M., M. Cole, K. Cook, B. Lafrance, M. Peters, J. Ransom, E. Sargent, V. Smoke, and A. Stairs. 2002. "Holistic Risk-Based Environmental Decision-Making: A Native Perspective." *Environmental Health Perspectives* 110: 259–64.

Arroyo, John. 2010. "Culture in Concrete: Art and the Re-imagination of the Los Angeles River as Civic Space." M.C.P. thesis, Massachusetts Institute of Technology. Accessed November 16, 2015.http://www.arroyoseco.org/arroyojcthesis.pdf.

Baird, George. 2015. *Writings on Architecture and the City*. London: Artifice.

Bannon, Bryan. 2009. "Developing Val Plumwood's Dialogical Ethical Ontology and Its Consequences for a Place-Based Ethic." *Ethics & The Environment* 14: 39–55.

———. 2011. "Flesh and Nature: Understanding Merleau-Ponty's Relational Ontology." *Research in Phenomenology* 41: 327–57.

———. 2014. *From Mastery to Mystery: A Phenomenological Foundation for an Environmental Ethic*. Athens: Ohio University Press.

———. 2015. "Resisting the Domination of Nature: Regarding Time as an Ethical Concept." *Environmental Philosophy* 11: 333–58.

Behringer, Wolfgang. 2010. *A Cultural History of Climate*. Malden: Polity.

Benjamin, Jessica. 1998. *The Shadow of the Other: Intersubjectivity and Gender in Psychoanalysis*. New York: Routledge.

Benner, Patricia, ed. 1994. *Interpretive Phenomenology: Embodiment, Caring, and Ethics in Health and Illness*. Thousand Oaks, CA: Sage.

Berleant, Arnold. 1992. *The Aesthetics of Environment*. Philadelphia: Temple University Press.

———. 2007. "Aesthetic Engagement and the Human Landscape." Presented at the international conference on Environment, Aesthetic Engagement, and the Public Sphere, Paris, May 9, 2007.

———. 2010a. *Sensibility and Sense: The Aesthetic Transformation of the Human World*. Exeter, UK: Imprint Academic.

———. 2010b. "Reconsidering Scenic Beauty." *Environmental Values* 19: 335–50.

Bielak, Alex, and Dana Mount. 2011. "Liquid Language: A Global Water Discourse." United Nations University. Accessed August 2015.http://unu.edu/publications/

articles/liquid-language-the-deepening-and-shallowing-of-global-water-discourse.html#info.

Biesecker, Adelheid, and Sabine Hofmeister. 2010. "Focus:(re) Productivity: Sustainable Relations Both Between Society and Nature and Between the Genders." *Ecological Economics* 69: 1703–11.

Biro, Andrew. 2005. *Denaturalizing Ecological Politics: Alienation from Nature from Rousseau to the Frankfurt School and Beyond.* Toronto: University of Toronto Press.

Blumenberg, Hans. 1985. *Work on Myth.* Translated by R. M. Wallace. Cambridge, MA: MIT Press.

Böhme, Gernot. 1989. *Für eine ökologische Naturästhetik.* Frankfurt: Suhrkamp.

Borrows, J. 2002. *Recovering Canada: The Resurgence of Indigenous Law.* Toronto: University of Toronto Press.

Bottici, Chiara. 2007. *A Philosophy of Political Myth.* Cambridge: Cambridge University Press.

———. 2008. "Mythos and Logos." *Epoché: A Journal for the History of Philosophy* 13: 1–24.

Brady, Emily. 2003. *Aesthetics of the Natural Environment.* Tuscaloosa: University of Alabama Press.

———. 2006. "Aesthetics in Practice: Valuing the Natural World." *Environmental Values* 15: 277–91.

Brand, Peg Zeglin, ed. 2000. *Beauty Matters.* Bloomington: Indiana University Press.

———, ed. 2013. *Beauty Unlimited.* Bloomington: Indiana University Press.

Briggle, Adam, Robert Frodeman, and Kelli Barr. 2015. "Achieving Escape Velocity: Breaking Free from the Impact Failure of Applied Philosophy." The London School of Economics and Political Science Impact Blog, April 27, 2015. Accessed July 2015.http://blogs.lse.ac.uk/impactofsocialsciences/2015/04/27/achieving-escape-velocity-applied-philosophy/.

Brown, Charles S. 2003. "The Real and the Good: Phenomenology and the Possibility of an Axiological Rationality." In *Eco-Phenomenology: Back to the Earth Itself,* edited by Charles S. Brown and Ted Toadvine, 3–18. Albany: State University of New York Press.

Brown, Valerie A., John A. Harris, and Jacqueline Y. Russell. 2010. *Tackling Wicked Problems: Through the Transdisciplinary Imagination.* London: Earthscan.

Bryant, L., N. Srnicek, and G. Harman, eds. 2011. *The Speculative Turn: Continental Materialism and Realism.* Melbourne: re-press.

Calhoun, A., M. Goeman, and M. Tsethlikai. 2007. "Achieving Gender Equity for American Indians." In *Handbook for Achieving Gender Equity through Education,* edited by S. S. Klein, B. Richardson, D. A. Grayson, L. H. Fox, C. Kramarae, D. S. Pollard, and C. A. Dwyer, 525–52. New York: Routledge.

Callicott, J. Baird. 1989. *In Defense of the Land Ethic: Essays in Environmental Philosophy.* Albany: State University of New York Press.

———. 1999. "Intrinsic Value in Nature: A Metaethical Analysis." In *Beyond the Land Ethic: More Essays in Environmental Philosophy.* Albany: State University of New York Press.

———. 2002. "Myth and Environmental Philosophy." In *Thinking Through Myths: Philosophical Perspectives,* edited by K. Schilbrack, 158–73. New York: Routledge.

———. 2011. "The Worldview Concept and Aldo Leopold's Project of 'World View' Remediation." *Journal for the Study of Religion, Nature and Culture* 5: 510–28.

Callicott, J. Baird, and Michael P. Nelson, eds. 1998. *The Great New Wilderness Debate.* Athens: University of Georgia Press.

Cameron, W. S. K. 2014. "Must Environmental Philosophy Relinquish the Concept of Nature? A Hermeneutic Reply to Steven Vogel." In *Interpreting Nature: The Emerging Field of Environmental Hermeneutics,* edited by Forrest Clingerman, Brian Treanor, Martin Drenthen, and David Utsler, 102–20. New York: Fordham University Press.

Capobianco, Richard. 2014. *Heidegger's Way of Being*. Toronto: University of Toronto Press.

Carlson, Allen. 2000. *Aesthetics and the Environment*: The Appreciation of Nature, Art and Architecture. London and New York: Routledge.

Carman, T. 2008a. "Review of T. Baldwin (ed.) *Reading Merleau-Ponty*: On Phenomenology of Perception." *Notre Dame Philosophical Reviews*.

———. 2008b. *Merleau-Ponty*. Abingdon: Routledge.

Casey, Edward. 1993. *Getting Back into Place: Toward a Renewed Understanding of the Place-World*. Bloomington: Indiana State University Press.

———. 2001. "Body, Self and Landscape: A Geophilosophical Inquiry into the Place-World." In *Textures of Place: Exploring Humanist Geographies*, edited by P. C. Adams et al., 403–25. Minneapolis: University of Minnesota Press.

City of Los Angeles. 2007. *The Los Angeles River Revitalization Master Plan*. Accessed November 16, 2015.http://boe.lacity.org/lariverrmp/CommunityOutreach/masterplan_download.htm.

———. 2015. "Oversight and Access." Accessed November 16, 2015. http://www.lariver.org/About/OversightandAccess/index.htm.

Clifford, Chadwick H. 2012. *Wilderness Rhythms: Playing Music to Enhance the Nature Experience*. Victoria, BC: Friesen.

Climate Change Science Program [CCSP]. 2009. *Thresholds of Climate Change in Ecosystems: A report by the US Climate Change Science Program and Subcommittee on Global Change Research*. Reston, VA: US Geological Survey.

Coates, K. 2004. *A Global History of Indigenous Peoples: Struggle and Survival*. New York: Palgrave Macmillan.

Coates, P. 1998. *Nature: Western Attitudes since Ancient Times*. Cambridge: Polity.

Conde, Marta. 2014. "Activism Mobilising Science." *Ecological Economics* 105: 67–77.

Coombes, B. L. 2013. "Maori and Environmental Justice: The Case of "Lake" Otara." In *Indigenous in the City: Contemporary Identities and Cultural Innovation*, edited by E. J. Peters and C. Andersen, 334–43. Vancouver: UBC Press.

Cooper, David E. 2012a. *Convergence with Nature: A Daoist Perspective*. Dartington: Green Books.

———. 2012b. "The Intimacy of Art and Nature," In *Human-Environment Relations*, edited by E. Brady and P. Phemister, 86–96. Dordrecht: Springer.

Coupe, Laurence. 1997. *Myth: The New Critical Idiom*. London and New York: Routledge.

Creswell, John W. 1998. *Qualitative Inquiry and Research Design: Choosing Among Five Traditions*. Thousand Oaks, CA: Sage.

Crist, E. 2004. "Against the Social Construction of Nature and Wilderness." *Environmental Ethics* 26: 5–24.

Crowell, Steven. 1996. "The Mythical and the Meaningless: Husserl and the Two Faces of Nature." In *Issues in Husserl's Ideas II*, edited by T. Nenon and L. Embree, 80–105. Dordrecht: Kluwer.

Daily, Gretchen C. 1997. *Nature's Services: Societal Dependence on Natural Ecosystems*. Washington, DC: Island Press.

Damasio, Antonio. 2012. *Self Comes to Mind: Constructing the Conscious Brain*. New York: Vintage.

Daniel, T. C. 2001. "Whither Scenic Beauty? Visual Landscape Quality Assessment in the 21st Century." *Landscape and Urban Planning* 54: 267–81.

Danto, A. C. 2003. *The Abuse of Beauty: Aesthetics and the Concept of Art*. Chicago: Open Court.

De Bary, Wm. Theodore, ed. 1960. *Sources of Chinese Tradition*. Vol I. New York: Columbia University Press.

Debussy, Claude. 1927. *M. Croche: The Dilettante Hater*. Translated by B. Langdon Davies. London: Viking.

Denzin, Norman K. 1992. "The Many Faces of Emotionality: Reading Persona." In *Investigating Subjectivity: Research on Lived Experience*, edited by Carolyn Ellis and Michael G. Flaherty, 17–30. London: Sage.

Derrida, Jacques. 2008. *The Animal That Therefore I Am*. Translated by David Wills. New York: Fordham University Press.

Descola, Philippe. 2011. "Human Natures." *Quaderns de l'institut Català d'Antropologia* 27: 11–16.

Detienne, Marcel, and Jean-Pierre Vernant. 1978. *Cunning Intelligence in Greek Culture and Society*. Translated by Janet Lloyd. Chicago: University of Chicago Press.

Dewey, John. 1958. *Experience and Nature*. New York: Dover.

———. 1980. *Art and Experience*. New York: Perigree.

Dewsbury, John David, and Paul Cloke. 2009. "Spiritual Landscapes: Existence, Performance and Immanence." *Social & Cultural Geography* 10: 695–711.

Diehm, Christian. 2008. "Staying True to Trees: A Specific Look at Anthropocentrism and Non-Anthropocentrism." *Environmental Philosophy* 5: 3–16.

Dillon, M. C. 1997. *Merleau-Ponty's Ontology*. 2nd ed. Evanston, IL: Northwestern University Press.

Dockry, M. J., K. Hall, W. Van Lopik, and C. M. Caldwell. 2015. "Sustainable Development Education, Practice, and Research: An Indigenous Model of Sustainable Development at the College of Menominee Nation, Keshena, WI, USA." *Sustainability Science* 11: 127–38.

Donohoe, Janet. 2014. *Remembering Places*. Lanham, MD: Lexington.

Dotson, K., and K. Whyte. 2013. "Environmental Justice, Unknowability and Unqualified Affectability. *Ethics & the Environment* 18: 55–79.

Drenthen, Martin. 2013. "New Nature Narratives: Landscape Hermeneutics and Environmental Ethics." In *Interpreting Nature: The Emerging Field of Environmental Hermeneutics*, edited by Forrest Clingerman, Brian Treanor, Martin Drenthen, and David Utsler, 225–44. New York: Fordham University Press.

Dreyfus, Hubert L., and Stuart E. Dreyfus. 1986. *Mind over Machine, The Power of Human Intuition and Expertise in the Era of the Computer*. Oxford: Basil Blackwell.

Dukes, Sheree. 1984. "Phenomenological Methodology in the Human Sciences." *Journal of Religion and Health* 23: 197–203.

Dunn, David. 2009. "Nature, Sound Art, and the Sacred." In *The Book of Music and Nature*, edited by David Rothenberg and Marta Ulvaeus, 95–107. Middletown, CT: Wesleyan University Press.

Ehrlich, Paul R., and Anne H. Ehrlich. 1981. *Extinction: The Causes and Consequences of the Disappearance of Species*. New York: Random House.

Embree, Lester. 1997. "Problems of the Value of Nature in Phenomenological Perspective or What to Do about Snakes in the Grass." In *Phenomenology of Values and Valuing*, edited by J. G. Hart and L. Embree, 49–61. Dordrecht: Kluwer Academic.

Embree, Lester, and Hans Rainer Sepp. 2010. "Introduction." In *Handbook of Phenomenological Aesthetics*, edited by Lester Embree and Hans Rainer Sepp, xv–xxx. Dordrecht: Springer.

Escobar, Arturo. 2008. *Territories of Difference: Place, Movements, Life, Redes*. Durham: Duke University Press.

Evernden, Neil. 1992. *The Social Creation of Nature*. Baltimore: Johns Hopkins University Press.

Feld, Steven. 2009. "Lift-up-over Sounding." In *The Book of Music and Nature*, edited by David Rothenberg and Marta Ulvaeus, 193–206. Middletown, CT: Wesleyan University Press.

Ferber, Marianne A., and Julie A. Nelson, eds. 1993. *Beyond Economic Man: Feminist Theory and Economics*. Chicago: University of Chicago Press.

Figueroa, R. M. 2006. "Evaluating Environmental Justice Claims." In *Forging Environmentalism: Justice, Livelihood, and Contested Environments*, edited by J. Bauer, 360–76. New York: M. E. Sharpe.

———. 2013. "Risking Recognition: New Assessment Strategies for Environmental Justice and American Indian Communities." *American Philosophcial Association Newsletter on Indigenous Philosophy* 12: 4–10.

Figueroa, R. M., and G. Waitt. 2008. "Cracks in the Mirror: (Un)covering the Moral Terrains of Environmental Justice at Uluṟu-Kata Tjuṯa National Park." *Ethics, Place and Environment* 11: 327–49.

Figueroa, R. M., G. Waitt, and L. McGee. 2007. "Fissures in the Rock: Rethinking Pride and Shame in the Moral Terrains of Uluṟu." *Transactions of the Institute of British Geographers* NS 32: 249–63.

Foltz, Bruce. 1995. *Inhabiting the Earth: Heidegger, Environmental Ethics and the Metaphysics of Nature*. Atlantic Highlands, NJ: Humanities.

———. 2004. "Nature's Other Side: The Demise of Nature and the Phenomenology of Givenness." In *Rethinking Nature: Essays in Environmental Philosophy*, edited by Bruce V. Foltz and Robert Frodeman, 330–41. Bloomington: Indiana University Press.

Frame, Bob, and Martin O' Connor. 2011. "Integrating Valuation and Deliberation: The Purposes of Sustainability Assessment." *Environmental Science & Policy* 14: 1–10.

Frodeman, Robert. 2014. *Sustainable Knowledge: A Theory of Interdisciplinarity*. New York: Palgrave Macmillan.

Fulton, William B. 2001. *The Reluctant Metropolis: The Politics of Urban Growth in Los Angeles*. Baltimore: Johns Hopkins University Press.

Gardiner, Stephen M. 2011. *A Perfect Moral Storm: The Ethical Tragedy of Climate Change*. New York: Oxford.

Gehlen, Arnold. 1988. *Man, His Nature and Place in the World*. Translated by C. McMillan and K. Pillemer. New York: Columbia University Press.

Geniusas, Saulius. 2012. *The Origins of the Horizon in Husserl's Phenomenology*. Dordrecht: Springer.

Georgescu-Roegen, Nicholas. 1971. *The Entropy Law and the Economic Process*. Cambridge, MA: Harvard University Press.

Gibson, James J. 1979. *The Ecological Approach to Visual Perception*. Hillsdale, NJ: Lawrence Erlbaum.

Gill, Jerry H. 1991. *Merleau-Ponty and Metaphor*. Atlantic Highlands, NJ and London: Humanities.

Giorgi, A. ed. 1985. *Phenomenology and Psychological Research*. Pittsburgh: Duquesne University Press.

Gómez-Baggethun, Erik, Rudolf De Groot, Pedros Lomas, and Carlos Montes. 2010. "The History of Ecosystem Services in Economic Theory and Practice: From Early Notions to Markets and Payment Schemes." *Ecological Economics* 69: 1209–18.

Gottlieb, Robert, and Andrea Misako Azuma. 2007. "Bankside Los Angeles." In *Rivertown: Rethinking Urban Rivers*, edited by Paul Stanton Kibel, 23–46. Cambridge, MA: MIT Press.

Griffiths, Paul. 2012. *Olivier Messiaen and the Music of Time*. London: Faber and Faber. Kindle Ed.

Grimley, Daniel M., ed. 2011a. *Jean Sibelius and his World*. Princeton, NJ: Princeton University Press. Kindle Ed.

———. 2011b. "Music, Landscape, Attunement: Listening to Sibelius." *Tapiola, Journal of the American Musicological Society* 64: 394–98.

Grinde, D. A., and B. E. Johansen. 1995. *Ecocide of Native America: Environmental Destruction of Indian Lands and Peoples*. Sante Fe, NM: Clear Light.

Grossman, Z., and A. Parker. 2012. *Asserting Native Resilience: Pacific Rim Indigenous Nations Face the Climate Crisis*. Corvalis: Oregon State University Press.

Gschwandtner, Christina M. 2014a. *Degrees of Givenness: On the Saturated Phenomenon of Jean-Luc Marion*. Bloomington: Indiana University Press.

———. 2014b. "Might Nature Be Interpreted as a 'Saturated Phenomenon'?" In *Interpreting Nature: The Emerging Field of Environmental Hermeneutics*, edited by Forrest

Clingerman, Brian Treanor, Martin Drenthen, and David Utsler, 82–101. New York: Fordham University Press.

Guha, Ramachandra. 2000. "The Paradox of Global Environmentalism." *Current History* 99: 367–70.

Guha, Ramachandra, and Juan Martínez Alier. 1997. *Varieties of Environmentalism: Essays North and South*. London: Earthscan Publications.

Gumprecht, Blake. 1999. *The Los Angeles River: Its Life, Death and Possible Rebirth*. Baltimore: Johns Hopkins University Press.

Habermas, Jürgen. 1973. *Legitimationsprobleme im Spätkapitalitimus.* Frankfurt: Suhrkamp Verlag.

Hall, Stuart. 1994. "Cultural Identity and Diaspora." In *Colonial Discourse and Postcolonial Theory: a Reader,* edited by Patrick Williams and Laura Chrisman, 392–401. London: Harvester Wheatsheaf.

Hapgood, Fred. 1993. *Up the Infinite Corridor: MIT and the Technical Imagination.* Reading, MA: Addison-Wesley.

Haraway, Donna. 1988. "Situated Knowledges: The Science Question in Feminism and the Privilege of Partial Perspective." *Feminist Studies* 14: 575–99.

Harðardóttir, María, et al. 2003. *Náttúruverndaráætlun 2004–2008—Aðferðafræði. Tillögur Umhverfisstofnunar um friðlýsingar* [The Nature Conservation Strategy 2004–2008. The Environment Agency's Suggestions for protection]. Reykjavík: Umhverfisstofnun [The Environment Agency of Iceland].

Harding, Sandra. 1991. *Whose Science? Whose Knowledge?: Thinking From Women's Lives.* Ithaca: Cornell University Press.

Harman, G. 2011. *Quentin Meillassoux: Philosophy in the Making*. Edinburgh: Edinburgh University Press.

Heft, Harry. 2003. "Affordances, Dynamic Experience, and the Challenge of Reification." *Ecological Psychology* 15: 149–80.

Hegel, G. W. F. 1975. *Aesthetics*. Translated by T. M. Knox. Oxford: Oxford University Press.

———. 1985. *Introduction to the Lectures on the History of Philosophy*. Translated by T. M. Knox and A. V. Miller. Oxford: Clarendon Press.

Heidegger, Martin. 1959. *An Introduction to Metaphysics.* Translated by Ralph Manheim. New Haven: Yale University Press.

———. 1969. *Discourse on Thinking.* Translated by J. Anderson and E. Freund. New York: Harper and Row.

———. 1973. *Being and Time.* Translated by John Macquarrie and Edward Robinson. Oxford: Blackwell.

———. 1977. *The Question Concerning Technology and other Essays.* Translated by William Lovitt. New York: Harper.

———. 1982. *The Basic Problems of Phenomenology*. Translated with an introduction by Albert Hofstadter. Bloomington: Indiana University Press.

———. 1991. *Nietzsche. Vol. 1: The Will to Power as Art*. Translated by D. F. Krell. San Francisco: HarperCollins.

———. 1993. *Basic Writings,* edited by D. F. Krell. London: Routledge.

———. 2002. "The Origin of the Work of Art." In *Off the Beaten Track,* edited and translated by Julian Young and Kenneth Haynes, 1–56. Cambridge: Cambridge University Press.

———. 2003. *Four Seminars.* Translated by Andrew Mitchell and François Raffoul. Bloomington: Indiana University Press.

Heller, Agnes. 2012. *The Concept of the Beautiful*. Edited and with an essay by Marcia Morgan. Lanham, MD: Lexington.

Hesiod. 1914. *The Theogony*. In *Hesiod, The Homeric Hymns and Homerica.* Translated by Hugh G. Evelyn-White. Cambridge, MA: Harvard University Press. Accessed November 16, 2015.http://data.perseus.org/citations/urn:cts:greekLit:tlg0020.tlg001.perseus-eng1:337.

Hinman, Lawrence M. 1982. "Nietzsche, Metaphor, and Truth." *Philosophy and Phenomenological Research* 43: 179–99.

Holland, Alan, and John O'Neill. 1994. "The Integrity of Nature over Time." In *The Thingmount Working Paper Series on the Philosophy of Conservation*. British Association of Nature Conservationists.

Homer. 1924. *The Iliad*. In *The Iliad with an English Translation*. Translated by Augustus Taber Murray. Cambridge, MA: Harvard University Press. Accessed November 16, 2015.http://data.perseus.org/citations/urn:cts:greekLit:tlg0012.tlg001.perseus-eng1:2.858.

Hoogeveen, D. 2015. "Sub-Surface Property, Free-Entry Mineral Staking and Settler Colonialism in Canada." *Antipode* 47: 121–38.

Hoover, E. 2013. "Cultural and Health Implications of Fish Advisories in a Native American Community." *Ecological Processes* 2: 1–12.

Hoover, E., K. Cook, R. Plain, K. Sanchez, V. Waghiyi, P. Miller, R. Dufault, C. Sislin, and D. O. Carpenter. 2012. "Indigenous Peoples of North America: Environmental Exposures and Reproductive Justice." *Environmental Health Perspectives* 120: 1645–49.

Hume, David. 1978. *A Treatise of Human Nature*. Oxford: Oxford University Press.

Hundley, Norris. 1992. *The Great Thirst: Californians and Water, 1770s–1990s*. Berkeley: University of California Press.

Husserl, Edmund. 1960. *Cartesian Meditations: An Introduction to Phenomenology*. Translated by Dorion Cairns. The Hague: Martinus Nijhoff.

———. 1970. *Crisis of the European Sciences and Transcendental Phenomenology*. Evanston, IL: Northwestern University Press.

———. 1981. "Foundational Investigations of the Phenomenological Origin of the Spatiality of Nature." Translated by Fred Kersten. In *Husserl: Shorter Works*, edited by Peter McCormick and Fred Elliston. Notre Dame, IN: University of Notre Dame Press.

———. 1983. *Ideas Pertaining to a Pure Phenomenology and a Phenomenological Philosophy*. Translated by F. Kersten. First Book. The Hague: Martinus Nijhoff.

———. 1998. *The Paris Lectures*. Translated by Peter Koestenbaum. Boston: Kluwer Academic.

———. 2001 [1900/1901]. Logical Investigations . 2nd ed. Edited by Dermot Moran. 2 vols. London: Routledge.

Ingold, Tim. 1993. "The Temporality of the Landscape." *World Archaeology* 25: 152–74.

———. 2011. *Being Alive*. New York: Routledge.

Intergovernmental Panel on Climate Change [IPCC]. 2014. *Climate Change 2014: Mitigation of Climate Change Contribution of Worlking Group III to the Fifth Assessment Report of the Intergovernmental Panel on Climate Change*. Cambridge: Cambridge University Press.

Jackson, Wes. 1994. *Becoming Native to this Place*. Berkeley: Counterpoint.

Jamieson, Dale. 2014. *Reason in a Dark Time: Why the Struggle Against Climate Change Failed*. New York: Oxford University Press.

James, Simon. 2009. *The Presence of Nature: A study in Phenomenology and Environmental Philosophy*. London: Palgrave Macmillan.

James, William. 1911. "Remarks at the Peace Banquet." In Memories and Studies, 299–306. New York: Longmans, Green.

Jantzen, Grace M. 2004. *Death and the Displacement of Beauty, Volume One: Foundations of Violence*. London and New York: Routledge Taylor & Francis Group.

Jax, Kurt, et al. 2013. "Ecosystem Services and Ethics." *Ecological Economics* 93: 260–68.

Johnson, Galen A. 2009. *The Retrieval of the Beautiful: Thinking Through Merleau-Ponty's Aesthetics*. Evanston, IL: Northwestern University Press.

Johnson, Stephen. 2012. *Mahler—His Life and Music*. Naxos Audiobooks, Kindle Ed.

Kant, Immanuel. 2002. *Critique of Practical Reason*. Indianapolis: Hackett.

Kassi, N. 1996. "A Legacy of Maldevelopment." In *Defending Mother Earth: Native American Perspectives on Environmental Justice*, edited by J. Weaver, 72–84. Maryknoll, NY: Orbis.

Keulartz, Jozef. 1998. *The Struggle for Nature: A Critique of Radical Ecology*. London: Routledge.

Kibel, Paul Stanton. 2007. "Bankside Urban: An Introduction." In *Rivertown: Rethinking Urban Rivers*, edited by Paul Stanton Kibel, 1–22. Cambridge, MA: MIT Press.

Kirwan, James. 1999. *Beauty*. Manchester: Manchester University Press.

Klaver, Irene J. 2005. "Always Already Rhythm, Merleau-Ponty's Movement Toward Ontology." *Studies in Practical Philosophy: A Journal of Ethical and Political Philosophy* 5: 41–49.

———. 2012. "Placing Water and Culture." In *Water, Cultural Diversity & Global Environmental Change: Emerging Trends, Sustainable Futures?*, edited by Barbara Rose Johnston, Lisa Hiwasaki, Irene J. Klaver, A. Ramos Castillo, and Veronica Strang, 9–29. The Hague: UNESCO International Hydrological Program.

———. 2013. "Environment Imagination Situation." In *Linking Ecology and Ethics for a Changing World: Values, Philosophy, and Action*, edited by Ricardo Rozzi, Steward T. Pickett, Clare Palmer, Juan J. Armesto, and J. Baird Callicott, 85–105. Dordrecht: Springer.

———. 2014a. "Landscapes of Environmental Imagination: Ranging from NASA and Cuyahoga Images to Kiefer and O'Keeffe Paintings." In *Environmental Aesthetics: Crossing Divides and Breaking Ground*, edited by Martin Drenthen and Jozef Keulartz, 135–53. New York: Fordham University Press.

———. 2014b. "Meander(ing) Multiplicity." In *Water Scarcity, Security and Democracy: A Mediterranean Mosaic*, edited by Francesca de Châtel, Gail Holst-Warhaft, and Tammo Steenhuis, 36–46. Ithaca: Cornell University Press, the Atkinson Center for a Sustainable Future, and Global Water Partnership-Mediterranean (GWP-Med).

Klaver, Irene J., and J. Aaron Frith. 2014. "A History of Los Angeles's Water Supply: Towards Reimagining the Los Angeles River." In *A History of Water, Series III, Vol. 1: Water and Urbanization*, edited by Terje Tvedt and Terje Oestigaard, 520–49. London: I. B. Tauris.

Kohák, Erazim. 2003. "An Understanding Heart: Reason, Value, and transcendental Phenomenology." In *Eco-Phenomenology: Back to the Earth Itself*, edited by Charles S. Brown and Ted Toadvine, 19–35. Albany: State University of New York Press.

Kolbert, Elizabeth. 2014. *The Sixth Extinction: An Unnatural History*. London: Bloomsbury.

Kosoy, Nicholas S., and Esteve Corbera. 2010. "Payments for Ecosystem Services as Commodity Fetishism." *Ecological Economics* 69: 1228–36.

Kövecses, Zoltán. 2010. *Metaphor: A Practical Introduction*. Oxford: Oxford University Press

Kruger, Dreyer. 1979. *An Introduction to Phenomenological Psychology*. Pittsburgh: Duquesne University Press.

Kuhn, Helmut. 1940. "The Phenomenological Concept of the 'Horizon.'" In *Philosophical Essays in Memory of Edmund Husserl*, edited by M. Farber, 106–24. Cambridge, MA: Harvard University Press.

Laduke, W. 1993. "A Society Based on Conquest Cannot Be Sustained." In *Toxic Struggles: The Theory and Practice of Environmental Justice*, edited by R. Hofrichter, 98–106. Philadelphia: New Society.

———. 1999. *All Our Relations: Native Struggles for Land and Life*. Cambridge, MA: South End Press.http://www.h-net.org/review/hrev-a0c0s9-aa.

Lakoff, George. 2010. "Why it Matters How We Frame the Environment." *Environmental Communication* 4: 70–81.

Lakoff, George, and Mark Johnson. 1999. *Philosophy in the Flesh: The Embodied Mind and its Challenge to Western Thought*. New York: Basic.

———. 1980. *Metaphors We Live By*. Chicago: University of Chicago Press.

Landau, Mark, Michael D. Robinson, and Brian P. Maier. 2013. *The Power of Metaphor: Examining its Influence on Social Life.* Washington, DC: American Psychological Association.

Landes, Donald A. 2013. *Merleau-Ponty and the Paradoxes of Expression.* London and New York: Bloomsbury.

Langer, Suzanne. 1957. *Philosophy in a New Key: A Study in the Symbolism of Reason, Rite and Art.* Cambridge, MA: Harvard University Press.

Larson, Brendan. 2011. *Metaphors for Environmental Sustainability: Redefining our Relationship with Nature.* New Haven: Yale University Press.

Latour, Bruno. 1993. *We Have Never Been Modern.* Translated by Catherine Porter. Cambridge, MA: Harvard University Press.

———. 1996. "On Actor-network Theory—A Few Clarifications." *Soziale Welt* 47: 369–81.

———. 2002. "Morality and Technology: The End of the Means." *Theory, Culture & Society* 19: 247–60.

———. 2004. *Politics of Nature: How to Bring the Sciences into Democracy.* Cambridge, MA: Harvard University Press.

———. 2013. "Facing Gaia: A New Enquiry into Natural Religion." Gifford Lecture. Accessed 2015, August 25. http://www.ed.ac.uk/schools-departments/humanities-soc-sci/news-events/lectures/gifford-lectures/archive/series-2012-2013/bruno-latour.

———. 2014. "Another Way to Compose the Common World." Paper presented at the The Ontological Turn in French Philosophical Anthropology—An Executive Session of the AAA Annual Meeting, Chicago, November 23.

Lefevre, T. A. 2015. "Settler Colonialism." In *Oxford Bibliographies in Anthropology,* edited by J. Jackson, 1–26. Oxford: Oxford University Press.

Leopold, Aldo. 1949. *A Sand County Almanac and Sketches Here and There.* New York: Oxford University Press.

Leung, Hok-Lin. 1992. *City Images.* Kingston, Ontario: Ronald P. Frye.

Lewis, C. S. 2013. *Studies in Words.* Cambridge: Cambridge University Press.

Llewelyn, John. 2004. *Seeing Through God: A Geophenomenology.* Bloomington and Indianapolis: Indiana University Press.

Lockwood, Julie L., Martha F. Hoopes, and Michael P. Marchetti. 2013. *Invasion Ecology.* 2nd Ed. Hoboken, NJ: Wiley Blackwell.

Macdonald, E., and S. Rang. 2007. "Exposing Canada's Chemical Valley: An Investigation of Cumulative Air Pollution Emissions in the Sarnia, Ontario Area." An ecojustice report.

Madison, G. B. 1973. *The Phenomenology of Merleau-Ponty: A Search for the Limits of Consciousness.* Athens: Ohio University Press.

———. 1990. "Flesh as Otherness." In *Ontology and Alterity in Merleau-Ponty,* edited by G. B. Johnson and M. B. Smith, 27–34. Evanston, IL: Northwestern University Press.

Madjar, Irena, and Jo Ann Walton, eds. 1999. *Nursing and the Experience of Illness: Phenomenology in Practice.* New York: Routledge.

Malpas, J. E. 1999. *Place and Experience: A Philosophical Topography.* Cambridge: Cambridge University Press.

Marcel, Gabriel. 2005. *Music and Philosophy.* Translated by S. Maddox and R. Wood. Milwaukee: Marquette University Press.

Marion, Jean-Luc. 2002a. *Being Given: Towards a Phenomenology of Givenness,* translated by Jeffery L. Kosky. Stanford: Stanford University Press.

———. 2002b. *In Excess: Studies of Saturated Phenomena.* Translated by Robyn Horner and Vincent Berraud. New York: Fordham University Press.

Martinez-Alier, Joan. 2009. "Social Metabolism, Ecological Distribution Conflicts, and Languages of Valuation." *Capitalism Nature Socialism* 20: 58–87.

Marx, Karl, and Friedrich Engels. 1974. *The German Ideology,* edited and introduced by C. J. Arthur. London: Lawrence and Wishart.

Marx, Werner. 1992. *Towards a Phenomenological Ethics: Ethos and the Life-World*. Albany: State University of New York Press.
Massey, Doreen. 2006. "Landscape as Provocation—Reflections on Moving Mountains." *Journal of Material Culture* 11: 33–48.
Mcgregor, D. 2009. "Honouring Our Relations: An Anishnaabe Perspective on Environmental Justice." In *Speaking for Ourselves: Environmental Justice in Canada*, edited by J. Agyeman, P. Cole, R. Haluza-Delay, and P. O'Riley, 27–41. Vancouver: University of British Columbia Press.
McMahon, Jennifer. 2011. "Towards a Unified Theory of Beauty." *Literature and Aesthetics* 9: 7–27.
McShane, Katie. 2007a. "Why Environmental Ethics Should Not Give Up on Intrinsic Value." *Environmental Ethics* 29: 43–61.
———. 2007b. "Anthropocentrism vs. Nonanthropocentrism: Why Should We Care?" *Environmental Values* 16: 169–85.
Meillassoux, Quentin. 2008. *After Finitude: An Essay on the Necessity of Contingency*. Translated by Ray Brassier. New York: Continuum.
Merchant, Carolyn. 1980. *The Death of Nature: Women, Ecology, and the Scientific Revolution*. New York: HarperCollins.
Merleau-Ponty, Maurice. 1962. *Phenomenology of Perception*. Translated by Colin Smith. London: Routledge and Kegan Paul.
———. 1964. *The Primacy of Perception and Other Essays on Phenomenological Psychology, the Philosophy of Art, History and Politics*. Translated by James M. Edie. Evanston, IL: Northwestern University Press.
———. 1968. *The Visible and the Invisible*. Edited by C. Lefort, translated by A. Lingis. Evanston, IL: Northwestern University Press.
———. 1993. "Eye and Mind," in *The Merleau-Ponty Aesthetics Reader: Philosophy and Painting*, ed. Galen A. Johnson, 121–50. Evanston, IL: Northwestern University Press.
———. 2002. *The World of Perception*. Translated by Oliver Davis, foreword by Stéphanie Ménasé, introduction by T. Baldwin. Abingdon: Routledge.
———. 2003. *Nature: Course Notes from the Collège de France*. Translated by Robert Vallier, compiled and with notes by Dominique Séglard. Evanston, IL: Northwestern University Press.
———. 2004. "Eye and Mind." In *Basic Writings*, edited by T. Baldwin, 290–324. London: Routledge.
———. 2012. *Phenomenology of Perception*. Translated by Donald A. Landes. New York: Routledge.
Mill, John Stuart. 1904. "On Nature." In *Nature, The Utility of Religion and Theism*. London: Watts & Co. for the Rationalist Press. An electronic version can be found athttp://www.lancaster.ac.uk/users/philosophy/texts/mill_on.htm.
Millennium Ecosystem Assessment. 2003. *Ecosystems and Human Well-being: A Framework for Assessment*. Washington, DC: Island Press.
———. 2005. *Ecosystems and Human Well-being: Synthesis*. Washington, DC: Island Press.
Mitchell, Andrew J. 2013. "The Fourfold." In *The Bloomsbury Companion to Heidegger*, edited by François Raffoul and Eric S. Nelson. London: Bloomsbury.
Mithen, Steven. 2006. *The Singing Neanderthals: The Origins of Music, Language, Mind and Body*. Cambridge, MA: Harvard University Press.
Mohai, P., D. Pellow, and J. T. Roberts. 2009. "Environmental Justice." *Annual Review of Environment and Resources* 34: 405–30.
Morris, D. 2013. "From the Nature of Meaning to a Phenomenological Refiguring of Nature." *Royal Institute of Philosophy Supplement* 72: 317–41.
Morton, Timothy. 2007. *Ecology without Nature: Rethinking Environmental Aesthetics*. Cambridge, MA: Harvard University Press.
Mothersill, Mary. 1984. *Beauty Restored*. Oxford: Clarendon Press.
Moustakas, Clark. 1994. *Phenomenological Research Methods*. Thousand Oaks, CA: Sage.

Mulholland, Catherine. 2000. *William Mulholland and the Rise of Los Angeles.* Berkeley: University of California Press.

Muraca, Barbara. 2007. "Getting over 'Nature': Modern Bifurcations, Postmodern Possibilities." In *Ecospirit: Religions and Philosophies for the Earth,* edited by Catherine Keller and Laurel Kearns, 156–77. New York: Fordham University Press.

———. 2010. *Denken im Grenzgebiet : Prozessphilosophische Grundlagen einer Theorie starker Nachhaltigkeit.* Freiburg/München: Alber.

———. 2011. "The Map of Moral Significance: A New Axiological Matrix for Environmental Ethics." *Environmental Values* 20: 375–96.

Nagel, T. 1986. *The View from Nowhere.* New York: Oxford University Press.

Neihardt, John. 2014. *Black Elk Speaks: The Complete Edition.* Lincoln: University of Nebraska Press.

Norberg-Schulz, Christian. 1979. *Genius Loci: Towards a Phenomenology of Architecture.* New York: Rizzoli.

Norgaard, Richard B. 2010. "Ecosystem Services: From Eye-Opening Metaphor to Complexity Blinder." *Ecological Economics* 69: 1219–27.

Noss, Reed F. 1998. "Sustainability and Wilderness." In *The Great New Wilderness Debate: An Expansive Collection of Writings Defining Wilderness from John Muir to Gary Snyder.* Edited by J. Baird Callicott and Michael P. Nelson, 408–13. Athens and London: University of Georgia Press.

O'Connor, Brian, Jud H. Copeland, and Jodi L. Kearns. 2003. *Hunting and Gathering on the Information Savanna.* Lanham, MD: Scarecrow.

O'Connor, Brian, and Roger B. Wyatt. 2004. *Photo Provocations, Thinking in, With, and About Photographs.* Lanham, MD: Scarecrow.

Orr, David. 1991. "What is Education For? Six Myths about the Foundations of Modern Education, and Six New Principles to Replace Them." In *The Learning Revolution* 27: 52. Accessed July 2015.http://www.context.org/iclib/ic27/orr/.

Petrarca, Francesco. 2006. *Petrarch's Ascent of Mount Ventoux: The Familiaris IV, Volume 23 of Scriptores Latini,* edited by R. J. Lokaj. Rome: Edizioni dell'Ateneo.

Plumwood, Val. 1993. *Feminism and the Mastery of Nature.* New York: Routledge.

———. 1998. "Wilderness Scepticism and Wilderness Dualism." In *The Great New Wilderness Debate: An Expansive Collection of Writings Defining Wilderness from John Muir to Gary Snyder.* Edited by J. Baird Callicott and Michael P. Nelson, 652–90. Athens and London: University of Georgia Press.

———. 2002. *Environmental Culture: The Ecological Crisis of Reason.* New York: Routledge.

———. 2005. "Toward a Progressive Naturalism." In *Recognizing the Autonomy of Nature: Theory and Practice,* edited by T. Heyd, 25–53. New York: Columbia University Press.

———. 2011. "Nature in the Active Voice." In *Climate Change and Philosophy: Transformational Possibilities,* edited by Ruth Irwin, 32–47. London: Continuum.

———. 2012. *The Eye of the Crocodile.* Edited by Lorraine Shannon. Canberra: Australian National University Press.

Polkinghorne, D. E. 1989. "Phenomenological Research Methods." In *Existential-phenomenological Perspectives in Psychology,* edited by R. S. Valle and S. Halling, 41–60. New York: Plenum.

Price, Jenny. 2008. "Remaking American Environmentalism: On the Banks of the L.A. River." *Environmental History* 13: 536–55.

Regan, Tom. 1981. "The Nature and Possibility of an Environmental Ethic." *Environmental Ethics* 3: 19–34.

Reisner, Marc. 1987. *Cadillac Desert: The American West and Its Disappearing Water.* New York: Penguin.

Reissmann, Erwin. 2012. "The Origin of the Meander." *blogmymaze.* Accessed November 16, 2015.http://blogmymaze.wordpress.com/2012/05/20/the-origin-of-the-meander/.

Ritter, Joachim. 1989. "Landschaft. Zur Funktion des Ästhetichen in der modernen Gesellschaft." In *Subjectivität*, 141–63. Frankfurt: Suhrkamp.
Robbins Landon, H. C. 1970. *Beethoven: A Documentary Study*. London: Thames & Hudson.
Rolston, Holmes III. 1982. "Are Values in Nature Subjective or Objective?" *Environmental Ethics* 4: 125–51.
———. 1988. *Environmental Ethics*. Philadelphia: Temple University Press.
———. 1997. "Nature for Real: Is Nature a Social Construct?" In *The Philosophy of the Environment*, edited by T. D. J. Chappell, 38–64. Edinburgh: Edinburgh University Press.
Rose, Philip. 2007. "Philosophy, Myth, and the 'Significance' of Speculative Thought." *Metaphilosophy* 38: 632–53.
Rothenberg, David, and Marta Ulvaeus. 2009. *The Book of Music and Nature*. Middletown, CT: Wesleyan University Press.
Russell, B. 1967. *The Problems of Philosophy*. Oxford: Oxford University Press.
Samuelsson, Lars. 2010. "On the Demarcation Problem and the Possibility of Environmental Ethics: A Refutation of 'A Refutation of Environmental Ethics.'" *Environmental Ethics* 32: 247–65.
Sarno, Louis. 1988. *Song from the Forest: My Life among the Ba-Benjellé Pygmies*. London: Bantam.
Sayes, Edwin. 2014. "Actor-network Theory and Methodology: Just What Does It Mean to Say that Nonhumans Have Agency?" *Social Studies of Science* 44: 134–49.
Scarry, Elaine. 2001. *On Beauty and Being Just*. Princeton, NJ: Princeton University Press.
Schafer, R. Murray. 2009. "Music and the Soundscape." In *The Book of Music and Nature*, edited by David Rothenberg and Marta Ulvaeus, 58–68. Middletown, CT: Wesleyan University Press.
Scruton, Roger. 2009. *Understanding Music: Philosophy and Interpretation*, London: Bloomsbury.
Seal, Jeremy. 2012. *Meander: East to West, Indirectly, Along a Turkish River*. New York: Bloomsbury.
Seamon, David. 2012. "Seeing and Animating the City: A Phenomenological Ecology of Natural and Built Worlds." In *The Natural City: Re-Envisioning the Built Environment*, edited by Ingrid Stefanovic and Stephen B. Scharper, 231–56. Toronto: University of Toronto Press.
Seamon, David, and Robert Mugerauer. 2000 [1989]. *Dwelling, Place and Environment: Toward a Phenomenology of Person and World*. Malabar, FL: Krieger [New York: Columbia University Press].
Sessions, G. 1995. "Postmodernism and Environmental Justice: The Demise of the Ecology Movement?" *The Trumpeter* 12. Accessed November 11, 2015. http://trumpeter.athabascau.ca/index.php/trumpet/article/view/305/456.
Shockley, K. 2012. "Human Values and Institutional Responses to Climate Change." In *Ethical Adaptation to Climate Change: Human Virtues of the Future*, edited by A. Thompson and J. Bendik-Keymer, 281–98. Cambridge, MA: MIT Press.
Shrader-Frechette, K. S. 2002. *Environmental Justice: Creating Equality, Reclaiming Democracy*. Oxford: Oxford University Press.
Singer, Peter. 2011. *Practical Ethics*. 3rd. ed. New York: Cambridge University Press.
Skúlason, Páll. 2005. *Meditations at the Edge of Askja*. Reykjavík: Háskólaútgáfan [University of Iceland Press].
———. 2008. "On the Spiritual Understanding of Nature." Paper presented by invitation of the Working Group on Religion, Ethics and Nature at Ohio Northern University, Ohio, April 15, 2008.
Small, Christopher. 1998. *Musicking: The Meanings of Performing and Listening*. Middletown, CT: Wesleyan University Press.
Smith, L. T. 1999. *Decolonizing Methodologies: Research and Indigenous Peoples*. London: Zed.

Smith, Stephen J. 2011. "Becoming Horse in the Duration of the Moment: The Trainer's Challenge." *Phenomenology & Practice* 5: 7–26.

Soper, Kate. 1995. *What is Nature?: Culture, Politics and the Non-Human*. Oxford: Blackwell.

Sparrow, T. 2014. *The End of Phenomenology: Metaphysics and the New Realism*. Edinburgh: Edinburgh University Press.

Spash, Clive L. 2008. "How Much Is that Ecosystem in the Window? The One with the Bio-diverse Trail." *Environmental Values* 17: 259–84.

Spiegelberg, Herbert. 1960. *The Phenomenological Movement*. 2 vols. The Hague: Martinus Nijhoff.

Spretnak, Charlene. 1991. *States of Grace: The Recovery of Meaning in the Postmodern Age*. New York: HarperCollins.

Stambovsky, Phillip. 2004. *Myth and the Limits of Reason*. Lanham, MD: University Press of America.

Stefanovic, Ingrid Leman. 1987. *Safeguarding Our Common Future: Rethinking Sustainable Development*. Albany: State University of New York Press.

———. 1991. "Evolving Sustainability: A Re-thinking of Ontological Foundations." *The Trumpeter* 8: 194–200.

Stefanovic, Ingrid Leman, and Stephen Scharper. 2012. *The Natural City: Re-Envisioning the Built Environment*. Toronto: University of Toronto Press.

Stein, Edith. 1989. *The Collected Works of Edith Stein: On the Problem of Empathy*. Vol. 3. Washington, DC: ICS.

Steinbock, Anthony. 1995. *Home and Beyond: Generative Phenomenology after Husserl*. Evanston, IL: Northwestern University Press.

———. 1998. "Spirit and Generativity: The Role and Contribution of the Phenomenologist in Hegel and Husserl." In *Alterity and Facticity: New Perspectives on Husserl*, edited by. N. Depraz and D. Zahavi, 163–203. Dordrecht: Kluwer.

———. 2003. "Generativity and the Scope of Generative Phenomenology." In *The New Husserl: A Critical Reader*, edited by D. Welton. Bloomington: Indiana University Press.

Strabo. 1877. *Geographica*. Leipzig: Teubner. Accessed November 16, 2015.http://www.perseus.tufts.edu/hopper/text?doc=Perseus:text:1999.01.0197.

———. 1924. *The Geography of Strabo*, edited by H. L. Jones. Cambridge, MA: Harvard University Press.

Takemitsu, Toru. 2009. "Nature and Music." In *The Book of Music and Nature*, edited by David Rothenberg and Marta Ulvaeus, 183–92. Middletown, CT: Wesleyan University Press.

Tannerbaum, Julie. 2015. "Mere Moral Failure." *Canadian Journal of Philosophy* 45: 58–84.

Tarbell, A. and M. Arquette. 2000. "Akwesasne: A Native American Community's Resistance to Cultural and Environmental Damage." In *Reclaiming the Environmental Debate: The Politics of Health in a Toxic Culture*, edited by R. Hofrichter, 93–112. Cambridge, MA: MIT Press.

Taylor, Steve J., and Robert Bogdan. 1998. *Introduction to Qualitative Research Methods: A Guidebook and Resource*. 3rd ed. New York: Wiley.

Tessman, Lisa. 2015. *Moral Failure: On the Impossible Demands of Morality*. New York: Oxford University Press.

Thoreau, Henry David. 1886. *Walden*. London: Scott.

Thorgeirsdóttir, Sigríður. 2005. "Heild sem gerir mann heilan: Um siðfræði náttúrufegurðar [A Wholeness that Makes You Whole: On the Ethics of Natural Beauty]," in *Hugsað með Páli* [Thinking with Páll—Festschrift for Páll Skúlason], edited by Róbert H. Haraldsson, Salvör Nordal, and Vilhjálmur Árnason, 163–79. Reykjavík: University of Iceland Press.

———. 2010. "Conversations with Ourselves in Metaphysical Experiences of Nature." In *Conversations with Landscapes*, edited by Karl Benediktsson and Katrín Anna Lund, 13–27. Burlington, VT: Ashgate.

Toadvine, Ted. 1999. "Naturalizing Phenomenology." *Philosophy Today* SPEP Supplement: 124–31.

———. 2009. *Merleau-Ponty's Philosophy of Nature*. Evanston, IL: Northwestern University Press.

———. 2014. "The Elemental Past." *Research in Phenomenology* 44: 262–79.

Tofighian, Omid. 2010. "Beyond the Myth/Philosophy Dichotomy: Foundations for an Interdependent Perspective." *Forum Philosophicum: International Journal for Philosophy* 15: 175–89.

Tuck, E., and R. A. Gaztambide-Fernández. 2013. "Curriculum, Replacement, and Settler Futurity." *Journal of Curriculum Theorizing* 29: 72–89.

Vogel, Steven. 2002. "Environmental Philosophy After the End of Nature." *Environmental Ethics* 24: 23–39.

———. 2015. *Thinking Like a Mall: Environmental Philosophy After the End of Nature*. Cambridge, MA: MIT Press.

Waage, Edda R. H. 2010. "Landscape as Conversation." In *Conversations with Landscapes*, edited by Karl Benediktsson and Katrín Anna Lund, 45–59. Burlington, VT: Ashgate.

———. 2012. "Landscape in the Sagas of Icelanders: The Concepts of Land and Landsleg." *Norsk Geografisk Tidsskrift* 66: 177–92.

Waage, Edda R. H., and Karl Benediktsson. 2010. "Performing Expertise: Landscape, Governmentality and Conservation Planning in Iceland." *Journal of Environmental Policy and Planning* 12: 1–22.

Walker, R., D. Natcher, and T. Jojola. 2013. *Reclaiming Indigenous Planning*. Montreal: McGill-Queen's Press.

Walters, K. L., S. A. Mohammed, T. Evans-Campbell, R. E. Beltrán, D. H. Chae, and B. Duran. 2011. "Bodies Don't Just Tell Stories, They Tell Histories." *Du Bois Review: Social Science Research on Race* 8: 179–89.

Warren, Karen. 2001. "Introduction to Part 3: Ecofeminism." In *Environmental Philosophy: From Animal Rights to Radical Ecology*. 3rd ed. Edited by Michael E. Zimmerman, 253–72. Upper Saddle River, NJ: Prentice Hall.

Waters, Colin, Jan A. Zalasiewicz, Michael A. Ellis, and Andrea M. Snelling. 2014. *A Stratigraphical Basis for the Anthropocene*. London: Geological Society.

Watkins, Holly. 2011. "Musical Ecologies of Place and Placelessness." *Journal of the American Musicological Society*, 64: 404–8.

Weaver, J. 1996. *Defending Mother Earth: Native American Perspectives on Environmental Justice*. Maryknoll, NY: Orbis.

Whitehead, Alfred North. 1948. *Science and the Modern World*. New York: Mentor.

———. 1967. *Science and the Modern World*. New York: Free Press.

Whyte, K. P. 2011. "Environmental Justice in Native America." *Environmental Justice* 4: 185–86.

———. 2013. "Justice Forward: Tribes, Climate Adaptation and Responsibility." *Climatic Change* 120: 117–30.

———. 2015. "Indigenous Food Systems, Environmental Justice, and Settler-Industrial States." In *Global Food, Global Justice*, edited by M. Rawlinson, 143–66. Newcastle upon Tyne, UK: Cambridge Scholars.

Whyte, K. P., M. J. Dockry, W. Baule, and D. Fellman. 2014. "Supporting Tribal Climate Change Adaptation Planning Through Community Participatory Strategic Foresight Scenario Development." In *Project Reports*, edited by D. Brown, W. Baule, L. Briley, and E. Gibbons. Ann Arbor, MI: Great Lakes Integrated Sciences and Assessments Center.

Wildcat, D. R. 2009. *Red Alert! Saving the Planet with Indigenous Knowledge*. Golden, CO: Fulcrum.

Wilde, Oscar. 1966. "The Decay of Lying." In *The Works of Oscar Wilde*, 970–92. London: Collins.

Wilkinson, C. F. 2005. *Blood Struggle: The Rise of Modern Indian Nations*. New York: Norton.

Williams, Raymond. 1985. *Keywords: A Vocabulary of Culture and Society*. New York: Oxford University Press.

Willett, Cynthia. 2014. *Interspecies Ethics*. New York: Columbia University Press.

Willox, A. C., S. L. Harper, V. L. Edge, K. Landman, K. Houle, and J. D. Ford. 2011. "The Land Enriches the Soul: On Climatic and Environmental Change, Affect, and Emotional Health and Well-Being in Rigolet, Nunatsiavut, Canada." *Emotion, Space and Society* 6: 14–24.

Wood, David. 2003. "What Is Eco-Phenomenology?" In *Eco-Phenomenology: Back to the Earth Itself*, edited by Charles S. Brown and Ted Toadvine, 211–33. Albany: State University of New York Press.

———. 2014. "My Place in the Sun." In *Interpreting Nature*, edited by Forrest Clingerman, Brian Treanor, Martin Drenthen, and David Utsler, 281–96. New York: Fordham University Press.

WWF. 2013. "The Economic Value of Ecosystem Services in the Lower Mekong." Accessed August 25 2015. http://d2ouvy59p0dg6k.cloudfront.net/downloads/report_economic_analysis_of_ecosystemservicesnov2013.pdf.

Wylie, John. 2007. *Landscape. Key Ideas in Geography*. London and New York: Routledge.

Yanagi, Sōetsu. 1989. *The Unknown Craftsman*. Tokyo: Kodansha International.

Zahavi, Dan. 2012. "Hið margslungna sjálf: Sjónarhorn reynslu og fræða [The Complex Self: Empirical and Theoretical Perspectives]." *Hugur* 24: 175–95.

———. 2014. *Self & Other: Exploring Subjectivity, Empathy, and Shame*. New York: Oxford University Press.

Zhuangzi. 2009. *The Essential Writings*. Translated by B. Ziporyn. Indianapolis: Hackett.

Ziporyn, B. 2009. "How the Tree Sees Me: Sentience and Insentience in Tiantai and Merleau-Ponty." In *Merleau-Ponty and Buddhism*, edited by Jin Y. Park and Gereon Kopf, 61–82. Lanham, MD: Rowman & Littlefield.

Index

About the Contributors

Bryan E. Bannon is an associate professor of philosophy and the director of the Environmental Studies and Sustainability Program at Merrimack College. He is the author of *From Mastery to Mystery: a Phenomenological Foundation for an Environmental Ethic* (2014).

David E. Cooper is professor of philosophy emeritus at Durham University, England. He has been a visiting professor at universities in America, Asia, Europe and Africa. His books include *Existentialism: A Reconstruction, World Philosophies: An Historical Introduction, A Philosophy of Gardens, Convergence with Nature: A Daoist Perspective* and *Sunlight on the Sea: Reflecting on Reflections.*

Janet Donohoe is currently dean of the honors college and professor of philosophy at the University of West Georgia. She is the author of several articles on phenomenology and place, as well as a new book titled *Remembering Places* (2014). She is the former book review editor for *Environmental Philosophy* and has served on the executive committee of the International Association for Environmental Philosophy.

Tom Greaves is senior lecturer in philosophy at the University of East Anglia, UK. He works predominantly on ecological phenomenology, especially the thought of Martin Heidegger, but his work encompasses environmental ethics, environmental aesthetics and the environmental humanities broadly conceived. He is director of the Environmental Philosophy Research Centre at UEA.

Simon P. James is a senior lecturer in philosophy at Durham University. He has written a number of articles and books on environmental philosophy, including *The Presence of Nature: A Study in Phenomenology and Environmental Philosophy* (2009). His book *Environmental Philosophy: An Introduction* was published in 2015.

Guðbjörg Rannveig Jóhannesdóttir is an adjunct lecturer at the Iceland Academy of Arts and a postdoctoral fellow at the University of Iceland's Institute for philosophy. She has written several articles on landscape, phenomenology and environmental aesthetics in Icelandic and international journals and a book chapter on landscape and aesthetic value in the essay collection *Conversations with Landscape*, edited by Karl Benediktsson and Katrín Anna Lund.

Irene J. Klaver is director of the Philosophy of Water Project www.water.unt.edu and professor in philosophy at the University of North Texas. Her research and teaching focus on sociopolitical and cultural dimensions of water. Klaver was UNESCO water and cultural diversity advisor (2008-2013) and coeditor of the UNESCO book *Water, Cultural Diversity & Global Environmental Change: Emerging Trends, Sustainable Futures* (2012). She made various documentaries on water issues, such as *River Planet* (2011) and *The New Frontier: Sustainable Ranching in the American West* (2010). From 2010 to 2014 Klaver served as co-director for the International Association of Environmental Philosophy.

Barbara Muraca is an assistant professor of philosophy at Oregon State University. Drawing from sustainability theory, environmental and social philosophy, and feminist philosophy, her research focuses more specifically on the philosophical foundations of political ecology and on global environmental justice. Currently, she is working on the role of concrete utopias and social experiments for a social ecological transformation. In particular, her research focuses on the Southern European 'Degrowth' movement and discourse. She also works on society-nature relations and how the understanding of our relation to what we call nature shapes our self-understanding as individuals and as communities.

Tim Christion Myers is a doctoral candidate in the departments of philosophy and environmental studies at the University of Oregon, where he is completing a dissertation on existential approaches to climate denial inspired by Marx, Heidegger and Merleau-Ponty. He has published in the journals *De Ethica* and *Environmental Philosophy*. In addition to ecophenomenology, his research interests include climate ethics, social and political theory, ecofeminism and the philosophy of nature.

Zachary Shefman is a candidate for the bachelor of laws and bachelor of civil law degrees at McGill University. From 2014 to 2015, he worked as a research assistant with Dr. Ingrid Leman Stefanovic on the project "Interpreting Interdisciplinarity: The Case of Environmental Studies".

Bryan Smyth is visiting assistant professor at the University of Mississippi. His research deals primarily with phenomenology and critical theory. He has recently published *Merleau-Ponty's Existential Phenomenology and the Realization of Philosophy* (2013), and is currently translating Merleau-Ponty's *Le monde sensible et le monde de l'expression*. He is also working on two book projects entitled *Hyperdialectical Materialism: Merleau-Ponty and the Political* and *Rethinking Reification*.

Elise Springer is an associate professor of philosophy at Wesleyan University. Her work in moral philosophy focuses on processes of moral change, criticism and moral dialogue. Thinking about the process of moral criticism has led her to think about moral responsibility in the context of a social ecology of attention to problems. She is the author of *Communicating Moral Concern* (2013).

Ingrid Leman Stefanovic is dean, faculty of environment, Simon Fraser University and professor emeritus, department of philosophy, University of Toronto. Her research explores how values and perceptions affect environmental decision making. Recent books include *Safeguarding Our Common Future: Rethinking Sustainable Development* and *The Natural City: Re-Envisioning the Built Environment*.

Mark Thorsby is a professor of philosophy and the chair of the Philosophy and Humanities Department at Lone Star College in Houston TX. He is currently a doctoral candidate in philosophy at the New School for Social Research. His work centers on the intersection between phenomenology and environmental ethics.

Kristina Welch is a PhD candidate in the School of Resource and Environmental Management and an instructor with the Centre for Sustainable Community Development at Simon Fraser University, Vancouver, BC. Her PhD work examines the role of gender in rural economic development. Welch obtained her B.Sc. from the University of Western Ontario in computer science and a master's in business administration from the University of British Columbia's Sauder School of Business.

Kyle Powys Whyte is Timnick Chair in the Humanities in the Department of Philosophy at Michigan State University. He is a faculty member of the Environmental Philosophy & Ethics graduate concentration and serves as a faculty affiliate of the American Indian Studies and Environmental Science & Policy programs. His primary research addresses moral and political issues concerning climate policy and indigenous peoples and the ethics of cooperative relationships between indigenous peoples and climate science organizations. He is an enrolled member of the Citizen Potawatomi Nation. He serves on the U.S. Department of Interior's Advisory Committee on Climate Change and Natural Resource Science and is involved in the Climate and Traditional Knowledges Workgroup, Michigan Environmental Justice Coalition, Everybody Eats: Cultivating Food Democracy, Humanities for the Environment, the Consortium for Socially Relevant Philosophy of/in Science and the American Philosophical Association Committee on the Status of Indigenous Philosophers. Whyte has received the K. Patricia Cross Future Leaders Award from the

Association of American Colleges and Universities (2009) and the Bunyan Bryant Award for Academic Excellence from Detroiters Working for Environmental Justice (2015).

www.ingramcontent.com/pod-product-compliance
Lightning Source LLC
LaVergne TN
LVHW101446200926
848483LV00001B/67